Modern Approaches to Chemical Reaction Searching

Modern Approaches to Chemical Reaction Searching

Proceedings of a Conference organised by the Chemical Structure Association at the University of York, England, 8-11 July 1985

Edited by Peter Willett

Gower

CHEMISTRY

Published by
Gower Publishing Company Limited
Gower House
Croft Road
Aldershot
Hants GU11 3HR
England

Gower Publishing Company
Old Post Road
Brookfield
Vermont 05036
USA

British Library Cataloguing in Publication Data

Modern approaches to chemical reaction searching : proceedings of a conference organised by the Chemical Structure Association at the University of York, England, July 1985.
1. Information storage and retrieval systems——Chemistry
I. Chemical Structure Association
II. Willett, Peter
025'.06541   Z699.5.C5

ISBN 0 566 03550 2

Printed in Great Britain by Blackmore Press, Shaftesbury, Dorset

Contents

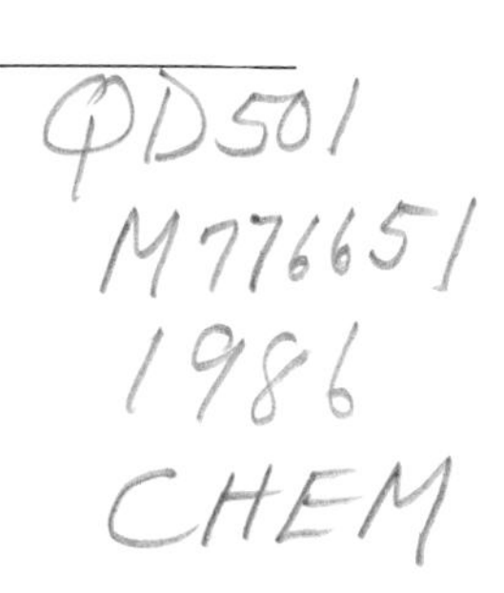

List of Authors viii

Introduction x

1. THE REACTION INDEXING PROBLEM: A HISTORICAL VIEWPOINT 1

   P. Willett

2. REACTION INFORMATION NEEDS OF THE SYNTHETIC CHEMIST 18

   D. P. J. Pearson

3. PRINTED TOOLS FOR REACTION SEARCHING 28

   I. Sinclair

4. THE CHEMICAL REACTIONS DOCUMENTATION SERVICE - PASTURES OLD AND NEW 36

   A. Finch

5. USER EXPERIENCE OF RETRIEVING CHEMICAL REACTION INFORMATION FROM PUBLICLY AVAILABLE ONLINE SERVICES 51

   P. T. Bysouth and J. Hardwick

6. EXPERIENCE WITH REACTION INDEXING AND SEARCHING IN THE IDC SYSTEM 68

C. Fricke, R. Fugmann, G. Kusemann, T. Nickelsen, G. Ploss and J. H. Winter

7. DESIGN, IMPLEMENTATION AND EVALUATION OF THE CONTRAST REACTION RETRIEVAL SYSTEM 78

D. Bawden and S. Wood

8. RMS-DARC. REACTION MANAGEMENT SYSTEM: A NEW SOFTWARE PRODUCED BY TÉLÉSYSTÈMES DARC 87

J. P. Gay

9. EXPLORING REACTIONS WITH REACCS 92

W. T. Wipke, J. Dill, D. Hounshell, T. Moock and D. Grier

10. SYNTHESIS LIBRARY 118

D. F. Chodosh

11. CREATION OF A CHEMICAL REACTION DATABASE FROM THE PRIMARY LITERATURE 146

P. E. Blower and R. C. Dana

12. THE BEILSTEIN-ONLINE DATABASE PROJECT 165

C. Jochum and S. Lawson

13. RETRIEVAL OF LITERATURE REFERENCES TO REACTIONS BY INPUT OF REACTANT AND PRODUCT STRUCTURES 170

M. Bersohn

14. AUTOMATIC KEYWORD GENERATION FOR REACTION SEARCHING 184

A. P. Johnson and A. P. Cook

15. DO WE STILL NEED A CLASSIFICATION OF REACTIONS? 202

G. Vladutz

16. A RECURSIVE REACTION GENERATOR 221

J. Brandt and K. Stadler

17. COMPOUND-ORIENTED AND REACTION-ORIENTED STRUCTURAL LANGUAGES FOR REACTION DATA BASE MANAGEMENT 240

J.-E. Dubois, G. Sicouri and R. Picchiottino

18. CONFERENCE OVERVIEW AND CLOSING REMARKS 257

W. A. Warr

19. Index 263

List of authors

D. Bawden and S. Wood
Pfizer Central Research,
Sandwich, Kent CT13 9NJ

M. Bersohn
Department of Chemistry, University of Toronto,
Toronto, Canada M5S 1A1

P. E. Blower and R. C. Dana
Chemical Abstracts Service, 2540 Olentangy River Road,
PO Box 3012, Columbus, Ohio 43210, USA

J. Brandt and K. Stadler
Technische Universität München,
D-8046 Garching, Lichtenbergstr. 4, West Germany

P. T. Bysouth and J. Hardwick
Glaxo Group Research,
Greenford, Middlesex

D. F. Chodosh
Smith, Kline and French Laboratories,
1500 Spring Garden Street, Philadelphia, PA 19101, USA

J.-E. Dubois, G. Sicouri and R. Picchiottino
Institut de Topologie et de Dynamique des Systemes,
Université Paris VII, 1 rue Guy de la Brosse, 75005 Paris

A. Finch
Derwent Publications Limited, Rochdale House,
128 Theobalds Road, London WC1X 8RP

C. Fricke, R. Fugmann, G. Kusemann, T. Nickelsen and
J. H. Winter
Hoechst AG, Postfach 80 03 20, D-6230 Frankfurt am Main 80,
West Germany

J. P. Gay
Telesystemes DARC, Tour Gamma B,
193-197 rue de Bercy, 75012 Paris

C. Jochum and S. Lawson
Beilstein-Institut, Varrentrappstr. 40-42,
D-6000 Frankfurt 90, West Germany

A. P. Johnson and A. P. Cook
Department of Organic Chemistry, University of Leeds,
Leeds LS2 9JT

D. P. J. Pearson
Imperial Chemical Industries PLC, Plant Protection Division,
Jealott's Hill Research Station, Bracknell, Berkshire
RG12 6EY

I. Sinclair
Pfizer Central Research
Sandwich, Kent CT13 9NJ

G. Vladutz
Institute for Scientific Information, 3501 Market Street,
University City Science Center, Philadelphia, PA 19104, USA

W. A. Warr
Imperial Chemical Industries PLC, Pharmaceuticals Division,
PO Box 25, Alderley Park, Macclesfield, Cheshire SK10 4TG

P. Willett
Department of Information Studies, University of Sheffield,
Western Bank, Sheffield S10 2TN

W. T. Wipke, J. Dill, D. Hounshell, T. Moock and D. Grier
Molecular Design Limited, 2132 Farallon Drive,
San Leandro, CA 94577, USA

Introduction

INTRODUCTION

Information pertaining to chemical reactions forms the basis for all synthetic planning in the laboratory and is thus of primary importance in both pure and applied chemistry. Despite this importance there have till recently been only a few tools to aid the chemist in his search for viable synthetic pathways. There are various reasons for this, for example the many characteristics of a reaction that may need to be noted for retrieval purposes, and these have meant that until recently, computerised systems for the handling of reaction information have been noticeably less advanced than comparable computerised systems for the handling of chemical structures.

The last few years have witnessed a considerable amount of research and development in the field of computerised reaction indexing and the Chemical Structure Association, in association with the Royal Society of Chemistry Chemical Information Group, organised an international conference, Modern Approaches to Chemical Reaction Searching, at the University of York from 8th to 11th July 1985 to review these developments and the current state-of-the-art in computerised reaction information systems.

The conference attracted some 125 delegates from around the world who are concerned with reaction information as users of chemical information systems, particularly those in the chemical and pharmaceutical industries, chemical database providers, and researchers in chemical structure handling. The conference included both formal presentations and extensive demonstrations of a range of computerised information systems. This volume is the formal proceedings of the conference and contains the texts of all but one of the nineteen papers that were presented, although their order has been altered slightly so as to allow a better presentation of the three main components of the conference. In essence these were:

* Current Systems and User Needs. In this section are papers describing the needs of practising chemists for reaction information, approaches to the problem of systematically indexing reactions for retrieval purposes, and currently available manual and computerised systems for the storage and retrieval of reaction information

* Computerised Systems. In this section are papers describing the new software systems for the handling of reaction information that have become available over the last few years. Examples are provided of systems that are available as commercial products, that have been designed specifically for use in-house by a single organisation, and that are used for the processing of public on-line data bases

* Research Topics in Reaction Information. In this section are papers providing an over-view of the research that is currently in-hand around the world into providing improved access to reaction information. The topics discussed include the generation of classification and coding schemes for reactions, algorithms for the processing and searching of reaction equations, and methods for integrating structural and textual data in reaction retrieval systems.

The paper by Willett ('The reaction indexing problem') introduces the section on Current Systems and User Needs, and provides a review of the many and varied approaches that have been suggested over the years for the analysis and indexing of chemical reactions. In particular, the paper provides a chronological over-view of much of the work that has been carried out at the University of Sheffield into automatic methods for the identification of the structural changes that characterise reactions. The other two groups that have made extensive studies of this problem are those at the Universities of Munich and Paris, and their work is described in the papers by Brandt and Stadler and by Dubois et al. respectively in the section Research Topics in Reaction Information.

The conference organising committee intended that the papers should cover a wide range of views of reaction information systems including not only information scientists but also chemists, the ultimate end-users of reaction information, and the next two papers are both by working chemists. Pearson ('Reaction indexing - a chemist's view') discusses the needs of the practising organic chemist for reaction information, with special reference to the various types of problem that are encountered by the synthetic chemist in his daily work. These problems are then used as the basis for a requirements specification that any effective reaction information system should be able to satisfy. Sinclair ('Printed tools for reaction searching') describes some of the many printed publications and services that chemists may need to consult when faced with a reaction information need.

The views of data base producers are represented by Finch ('The Chemical Reactions Documentation Service - pastures old and new') who describes the Chemical Reactions Documentation Service. This is produced by Derwent Publications, and is one of the best established current sources of reaction information, having been in operation for some ten years and having a database that derives from Theilheimer's renowned yearbooks Synthetic Methods of Organic Chemistry and from the Journal of Synthetic Methods.

Bysouth and Hardwick ('User experience of retrieving chemical reaction information from publicly available online services') give a critical review of the reaction files currently available online, and report the results of comparative searches on several such files. The paper is particularly useful since it gives an extended list of the features and facilities that are required of a reaction information system by the information scientist, and thus usefully complements the list of features provided by Pearson from the view of the synthetic chemist.

The section on Computerised Systems is the longest of the three sections of the book. The seven papers can be subdivided into those systems that have been developed by a single organisation for their own use in-house, systems that are commercially available for use in-house, and systems that are designed for public access.

Fricke et al. ('Experience with reaction indexing and searching in the IDC system') describe the reaction documentation system that has been developed by the companies of Internationale Dokumentationsgesellschaft für Chemie mbH. The system is based upon the well-known GREMAS fragmentation code that was originally developed for the indexing and searching of structures. In this respect, the system is comparable to that described in the paper by Bawden ('The CONTRAST reaction retrieval system') who discusses the design and implementation of an in-house reaction information system which has been developed at Pfizer U.K. and which again draws strongly upon an existing structure information system. The two papers differ, however, in the way that the reaction site indexing is performed, with the IDC system involving manual coding whereas the Pfizer system is based on a fully algorithmic approach to reaction site detection. Bawden's paper also describes the attempts that have been made to integrate the reaction system with other components of the Pfizer chemical information system: there seems little doubt that reaction information systems can only be utilised fully if such an integration takes place.

The papers by Gay ('RMS-DARC. Reaction management system: a new software produced by Telesystemes DARC') and Wipke et al. ('Exploring reactions with REACCS') describe two commercial software packages that have been developed for the in-house storage and retrieval of chemical reaction information. Both packages involve the use of computer graphics as the primary means of providing a user-friendly interface to reaction data, and allow a wide range of query types to be handled. There are, of course, differences between the two approaches, but the similarities help to emphasise the fact that there is now a well-understood core of reaction handling techniques that can be used to provide efficient and effective access to reaction data. The graphics-based system described in the paper by Chodosh ('SYNthesis LIBrary') is rather different in concept from the other in-house systems discussed in the conference since it has been designed specifically as a browsing aid for synthetic chemists, rather than as a precision retrieval tool that can be used by the information scientist intermediary. In this respect it is in many ways complementary to, rather than being a competitor of, the other packages that are commercially available.

Blower and Dana ('Creation of a chemical reaction database from the primary literature') discuss the routines and procedures that are under development at Chemical Abstracts Service for the creation of a database of chemical reactions that is to become available as a public on-line file in the near future. The paper gives an account of some of the design decisions that need to be made when a public database is set up. A feature of this database is that it is to be fully comprehensive in its coverage of the primary literature, whereas all of the other files of reactions that were described at the Conference have been carefully selected on grounds of novelty or perceived utility. Jochum and Lawson ('The Beilstein online database project') give an overview of the development programme that is currently underway to put the Beilstein Handbook of Organic Chemistry into machine-readable form. The paper does not deal specifically with reaction searching, but the huge range of validated information in this database, which has abstracted the primary literature since 1830, means that it will become a prime source of reaction information when the online project is completed.

Although the packages and projects described in the Computerised Systems section clearly provide sophisticated means of access to reaction data there are still several areas where research and development work is needed to further improve and extend system facilities: this work forms the basis for the papers in the section Research Topics in Reaction Information.

Two papers consider the use of knowledge engineering techniques in reaction information systems. Bersohn ('The retrieval of literature references to reactions by input of reactant and product structures') describes a computer program for reaction retrieval that has as its basis some of the techniques that have been developed for computer-aided synthesis design. In particular, the program is knowledgeable about features such as stereochemistry and conflicting functionality, which may not be included in general-purpose reaction retrieval packages. Johnson and Cook ('Automatic keyword generation for reaction searching') note the importance of keywords for characterising reactions and describe an automatic means for generating and maintaining a thesaurus of such descriptors: this is being implemented as a rule-based system using PROLOG.

Vladutz ('Do we still need a classification for reactions?') provides perhaps the most forward-looking of the papers in suggesting a browsing approach to the retrieval of chemical reaction information. This involves the matching of a query, expressed as a substructural transformation, against a file of automatically indexed reactions with some measure of similarity being calculated so as to identify the reactions in the file that are most similar to the query reaction. The paper by Brandt and Stadler ('A recursive reaction generator') describes a machine representation for the documentation of chemical reactions that can also be used for the generation of candidate product molecules, given a suitable set of starting materials. Such a system could form the basis for an intelligent front-end to a reaction information system, and again emphasises the close relationship between reaction indexing and synthesis planning. Dubois _et al_. ('Compound-oriented and reaction-oriented structural languages for reaction data base management') compare two coding schemes for reactions, one of which is designed to handle reactant-product molecular pair data while the other considers just the transformation(s) that take place during the course of a reaction. Theoretical and practical results are presented to demonstrate the superiority of the latter model.

The book is completed by Warr's final paper at the actual conference ('Conference overview and closing remarks') which gives her individual response to the papers that had been presented.

I am grateful to the authors for submitting their papers so as to allow the rapid publication of this book, to Virginia Messenger for her careful typing of the manuscripts, to Sue McNaughton for arranging things at Gower Publications Ltd, and to Digital Equipment Corporation for the provision of hardware during the conference. Finally, I would like to thank Peter Bamfield, Josef Brandt, Peter Johnson, John Rayner, Linda Richards, Bill Town, George Vladutz, Wendy Warr and Todd Wipke, my colleagues on the Organising Committee, for their contributions to the smooth running of the conference.

Peter Willett
Sheffield, October 1985

1 The reaction indexing problem: a historical viewpoint

P. Willett
University of Sheffield

INTRODUCTION

The advent of the computer has led to rapid changes in the methods that are used for the analysis, storage and retrieval of (primarily) scientific and technical information. Nowhere is this more true than with information pertaining to chemical structures where the availability of machine-readable structure representations, such as Wiswesser Line Notation (WLN) or connection tables of various sorts, has led to the development of a wide range of sophisticated search techniques, as well as systems for structure-activity prediction, structure elucidation and the like (Ash *et al.*, 1985).

The systems described in this book all relate to the provision of rapid and easy access to chemical reaction information. This is, of course, of fundamental importance to many branches of chemistry, but we shall be concerned here with the field of small-molecule synthetic organic chemistry where the need for adequate means of retrieval has been apparent for many years now. Thus the preface of the first edition of Weyl's *Die Methoden der Organische Chemie* (Weyl, 1901) contained the statement that a scientist could hardly hope to be familiar with every one of the innumerable methods described therein. More recently, both Kresze (1970) and Valls (1973) have called attention to the importance of providing adequate reaction information, while it has been stated that approximately one half of all the organic queries put by chemists to the BASF Ludwigshafen Documentation Group were concerned with reactions (Meyer, 1975). As there are now over seven million compounds known and any one may be transformed into many others by suitable reactions, it can be seen that the amount of potential data is enormous. Not only is the size of the potential data base constantly increasing, but it has also been pointed out that there are large classes of reactions for which there are as yet no known members (Hendrickson, 1975). There are often many ways in which a molecule may be synthesised and yet there are currently few aids to help the chemist in his search for a viable synthetic pathway. The difficulty of the problems involved may be evidenced both by the wide recognition of the achievements of chemists such as Corey and Woodward and by the frequent use of terms such as 'elegant' in reviews of syntheses. Synthetic organic chemistry has indeed been described as 'an art in the midst of a science' (Hendrickson, 1976).

It might have been expected that computers, with their ability to compare and collate large volumes of data would provide a ready means for the control of chemical reaction data but this has not proved to be so. At least in part, this lack of success has been due to the limited amount of research carried out in the field since the documentation of a reaction presupposes a method for the encoding of the reacting molecules, or some portion of them, which has only become feasible within the last ten years or so. There are, however, two major problems that need to be overcome before retrieval systems for reactions become as well-established as comparable systems for the storage and retrieval of individual molecules.

The first of these problems is the very wide range of types of query that a reactions retrieval system will need to handle. In order of increasing complexity, these include:

* Reactions in which both the reactant and product molecules are fully specified
* Reactions that give rise to some specific product molecule or substructure
* Reactions of a specific molecule or substructure
* Substructural transformations in which only the reactant and product substructures are specified

Methods for structure searching (Ash et al., 1985) may obviously be used to encompass the first three classes of query, while substructure search techniques may be used for the second and third classes, although there may be a lack of search precision if it is not possible to identify whether the substructure in question has actually been involved in the reaction or merely occurs elsewhere in the molecule. However such retrieval mechanisms are quite inapplicable to the last class, which is perhaps the most common sort of reaction query. In addition to the structural queries listed above, chemists may also be interested in other features of a reaction such as:

* Reactions involving a given reagent

* The effect of pressure upon the yield of a reaction

* Reactions used by a specific synthetic chemist

This range of query types leads one directly to the second main problem facing the designer of a reactions retrieval system viz the fact that whereas a chemical molecule is a unique entity, and thus susceptible to listing in some canonical form, such as a CAS Registry Number, a reaction has very many parts, all of which may need to be stored for subsequent retrieval. The large number of characteristics makes the organisation of the data quite a difficult task as will be clear from the following list of data elements, all of which at least should be available for search purposes:

* Compound information, this ideally including not just the starting materials and end-products but also any intermediates that are formed en route

* Experimental conditions such as added reagents, the temperature and pressure, and the yield of the reaction

* Citation details, such as a reference to a journal article or to a laboratory notebook

* Reaction analysis, that is some description of the change that has taken place

The reaction analysis is the most important and also least well defined data element: the lack of an efficient and effective means of analysing and indexing reactions has been the main reason why the development of reaction retrieval systems has lagged far behind the comparable systems for the storage and retrieval of individual structures and substructures.

This chapter presents an over-view of the many and varied methods that have been suggested for the indexing of chemical reactions, thus providing the historical background to the new systems and services that are described in the remainder of the book. It should be noted that only the most important methods are described here, and the reader is referred to Willett (1978) for a more extended survey of indexing approaches.

MANUAL METHODS FOR THE INDEXING OF REACTIONS

This section describes reaction indexing methods where the intellectual tasks of analysis and representation have been performed manually, albeit for subsequent mechanised storage and retrieval in some cases.

As with compound information, the earliest forms of reaction indexing were based upon nomenclature and to this day the most widely employed and most easily understood description is the use of a trivial name, usually that of the chemist(s) who originally discovered the reaction. Terms such as Skraup synthesis, Claisen condensation and Clemmensen reduction are common in the literature and several compendia are available, the most comprehensive of these containing several hundreds of entries. Nomenclature may occasionally prove very powerful in rapidly describing complexes which can be difficult to characterise using more systematic methods e.g., the Cope rearrangement. Generally, however, the use of indexing terms which have no direct relationship with the reaction that they are supposed to describe may lead to severe problems in retrieval (Fugmann et al., 1979). Thus structurally similar transformations may be separated which might be considered more fruitfully in conjunction and there may also be disagreement as to the exact extent of the reactions that should be considered under a single heading. However, the greatest deficiency is simply the lack of coverage offered by such a system since the overwhelming number of reactions have not been graced by a suitable appellation.

A more systematic use of nomenclature has been suggested by Patterson and Bunnett (1954) who proposed that the name of a substitution reaction should be composed of the name of the incoming group, the syllable 'de', the name of the outgoing group and the suffix 'ation': thus the hydrolysis of an alkyl chloride would be called hydroxydechlorination. The International Union of Pure and Applied Chemistry (IUPAC) has shown interest in the extension of the scheme (Ash et al., 1985) although Vleduts (1963) has highlighted ambiguities even in the case of simple functional group interconversion reactions.

An example of a reaction publication that is based upon nomenclature is Organic Syntheses which is devoted to the description of preparations of specific compounds so that the indexing is primarily upon the basis of the name of the product. Mischenko has described an index to the Russian translation of this publication in which broad classes, such as halogenation or nitration, are subdivided by a structural expression of the particular reaction class (1971). Cohen (1982) describes a reaction catalogue in which reactions are listed upon the basis of structural codes characterising the product molecules, sub-divided by the corresponding reactant codes.

It is convenient at this point to mention the use of indexes of functional groups and of reagents. The former are usually arranged by the functional group of the product and then subdivided by the functional group of the reactant which has been involved in the change. Obviously, such an approach can only deal satisfactorily with simple changes, especially if the reacting molecules are polyfunctional. Examples of reagent based indexes are Synthetica Merck and the well known Fieser and Fieser: under each reagent is listed the types of reactions for which it may be employed, usually with details of the appropriate reaction conditions.

A more systematic approach is to classify reactions according to the bonds broken or formed in the course of the reaction, an idea first proposed by Weygand (1938). Theilheimer developed Weygand's system to produce a simple classification based on the types of bonds broken and formed and on the nature of the reaction: this classification forms the basis for the series *Synthetic Methods of Organic Chemistry*. Reactions are described by a three part symbol string; the first part refers to the bond formed in the reaction, the second is a bond change indicator and the third the bond broken. The indicators represent addition, rearrangement, exchange and elimination reactions though these terms are used in a very broad sense. Further subdivision is possible on the basis of the reagents but this is not included in the symbol string. When a reaction involves more than one bond change, multiple entries are supposed to be made although this does not always appear to occur and one also finds that the set of reactions denoted by a single symbol string often bear little relationship to one another: both of these points are discussed at some length by Vleduts (1963), who also points out that it is often easier to find a reaction *via* the subject index rather than *via* the bond classification. The French firm Roussel-Uclaf operate a card file based on bond formation and reacting group data (Valls, 1973) and this method of classification has been employed in the series *Cahiers de Synthese Organique*. Bond change data is also included in the *Chemical Reactions Documentation Service* (CDRS) which is described in detail by Finch later in this book.

The most fruitful development of Weygand's idea has been the concept of the reaction centre, or reaction site, which seems to have been first described by Vleduts (1963). In his paper, he advocated the use of all the bond changes occurring during the reaction, rather than the single changes considered by Theilheimer. As he points out 'a distinctive feature of organic reactions, which involve complicated molecules containing almost exclusively covalent bonds, is the destruction and creation of a comparatively small number of bonds in such a way that, during the process, fairly extensive portions of the molecules do not change their structures.' This being so, we may attempt to classify reaction information upon the basis of the bonds that have been altered in the course of the reaction; taken together, these bonds represent the partial structures involved in the change, the reaction centre. To quote again, 'the essence of the work in developing a skeleton scheme of a particular reaction lies in the comparison of the structure of the final and initial molecules and in discarding the fragments of the structure not undergoing changes in the course of the reaction.' Such a skeletal reaction scheme will generally represent several similar reaction types since groups adjacent to the reaction sites are omitted although they may play a significant part in determining the course of the reaction in terms of yield, stereochemistry and overall structural change. The neglect of the nonreacting parts of the molecules is claimed as an advantage (Vleduts, 1963; Ziegler, 1966) since supposedly useful analogies may be detected between different reactions belonging to the same basic class: however, there are no generally available guidelines as to exactly what should be included in the reaction centre. Vleduts suggested that the site should consist of all the bonds altered during the reaction plus the following:

* any heteroatoms that are directly connected to an atom in the reaction site (a key atom)

* any atoms connected by multiple bonds to a key atom

* any groups of the form A=B or A≡B where A and B are any atoms of which at least one is attached to a key atom.

It may be noted that this selection of 'activating groups' is made upon a structural basis rather than upon the basis of any mechanistic considerations; moreover, such groups can be detected algorithmically with relatively little effort whereas the identification of the actual activating substructures implies a high degree of machine intelligence and significantly greater computational requirements. A greatly extended list of features has been described by Bersohn and Esack (1976).

Ziegler, in the Reactiones Organicae, has produced a set of punched cards embodying the reaction site concept (Ziegler, 1966; Ziegler, 1979). Each card bears a skeletal reaction scheme and the structure of the product, these being described by a simple fragmentation code, as well as a printed abstract and additional physical information such as conditions and neighbouring groups. The advantages cited by Ziegler are:

* a precise definition of the reaction type, independent of generic types such as oxidation

* easy detection of analogous reactions as only the reacting parts of the molecules are coded

* no assumptions are made as to the mechanism of the reaction

* the use of traditional symbols since the skeletal scheme is printed upon the card as well as being punched for machine use

* independent of nomenclature

* easy linking of the reaction centre with the whole molecule

* easy classification of reactions.

It seems clear that a reaction centre approach holds distinct promise and it has played a large part in the fully automatic indexing procedures described in the next section of this chapter. The Pharma system, which forms the basis of Derwent's CRDS and is prepared manually, has a limited amount of reaction centre information and an experimental reaction file at ICI Pharmaceuticals, again based on computer processing of manual input, was based entirely on the reaction centre approach (Eakin and Hyde, 1973).

Two widely used reaction documentation services are those developed by the Internationale Dokumentationsgesellschaft für Chemie (IDC) (Fugmann et al., 1979) and by the Pharma Documentation Ring (Valls and Schier, 1975), two consortia of European chemical and pharmaceutical firms. Both systems employ manually assigned fragmentation codes which are stored for subsequent machine search. In the GREMAS code of IDC each carbon atom is coded by at least one term consisting of three letters and a reaction is described by pairs of these terms corresponding to the initial and final states of every functional carbon atom modified in the course of the reaction. Various subsidiary terms are used to indicate the general type of the reaction, e.g., chain elongation or ring closure, and a variety of search techniques is available. The Pharma service is based on the fragmentation code of Derwent's RINGDOC patent alerting service and a limited amount of bond change and condition data are also included.

In closing, one should note that the great triumph of physical organic chemistry over the last thirty years or so has been the development of mechanistic theory by which it is possible to rationalise many known reactions upon the basis of inter- and intra-molecular electronic effects, and a comparable degree of coverage could presumably be achieved in the documentation area by employing some sort of mechanism based indexing. Qualitative descriptions of reaction mechanisms have been suggested (Guthrie, 1975; Satchell, 1977) but a quantitative description could only be achieved by wave mechanics equations, these being pictorially represented on a reaction diagram by electron shifts, charge transfer complexes and the like. This is another area that is under study by IUPAC at present (Ash et al., 1985).

Although the systems described in this section may be effective in operation, there are considerable problems of efficiency associated with their use. This is because reaction analyses are carried out purely by manual means, resulting in time-consuming and expensive systems that cannot be applied on a routine basis to files containing thousands or tens of thousands of reactions. Accordingly, there has been a considerable amount of interest in the development of reaction indexing systems that could perform reaction analysis by fully automated methods. This work, the bulk of which has been carried out in the University of Sheffield by Lynch and his co-workers, is described in the following section.

AUTOMATIC METHODS FOR INDEXING OF REACTIONS

The earliest suggestion that reaction analysis could be carried out automatically was made by Vleduts (1963) and shortly afterwards, Mischenko et al. (1965) reported an algorithm by which this might be performed. The underlying assumption was made that the bonds formed in the reaction would be different from those destroyed; thus a simple comparison of the bonds in the reactant and product molecules would reveal those that had changed. The input to the program consisted of the redundant connection tables of the reacting molecules and these were used to generate the lists of reactant and product bonds, the bond representatives consisting of the component atoms plus the bond order. Bonds common to the two sides of the equation were deleted and the remaining bonds were used as the basis for binary descriptors in a punched card retrieval system. Analyses were produced for 85 per cent of a sample file of ten thousand reactions and of those analysed, circa 75 per cent were judged as being correct.

A sophisticated development of this work was reported by Harrison and Lynch (1970). The differences between the reactant and product structures were again determined by analysing the two sets of reacting molecules into small bond-centred fragments which were then compared with each other to identify the structural differences. If the pair analysis was successful in detecting some structural differences the fragments were joined together to form a skeletal reaction scheme. The assembly of the reaction sites, which was carried out separately for each half of the equation, was essentially the construction of a partial connection table record of the reaction, the assembly being carried out in much the same way as one might construct a jigsaw puzzle with the parent tables acting as a sort of template for the rebuilding, and with the most compact reaction site(s) being constructed if an alternative were possible. Once the sites had been generated and validated, they, or rather the partial connection tables that represented them, were compacted for storage and written to an output file.

An extensive series of evaluations of this approach was carried out, as summarised by Willett (1978), and showed that the algorithm provided a relatively efficient and effective means of identifying reaction sites automatically. However, severe difficulties were revealed when the reaction sites were used as the basis for a simple, screen-based retrieval system. In part these limitations arose from the rather rudimentary fragment screens that were used, but the main problem was that there was no relationship between the reaction centres and the remaining, unchanged parts of the reacting molecules. It was concluded that a successful reactions retrieval system would need to allow searches to be carried out upon not only the reaction centres but also upon the unchanged reactant and product substructures, the reaction sites merely being represented by a suitable annotation of the appropriate atoms in the connection tables of the reacting molecules.

At about the same time, work was being carried out using WLN records for the reacting molecules. There are difficulties associated with the use of such representations owing to the lack of explicit connectivity information and to the fact that a few WLN symbols may represent quite large numbers of atoms and bonds, this implying that the analysis may be described in rather broad terms in some cases. The advantages of a WLN-based system are three-fold:

* it is easy to identify the (sub)structure corresponding to a WLN symbol string, thus opening the way to the production of KWIC-like printed indexes of reactions

* the notation gives special prominence to ring systems and functional groups and thus relatively simple programs should suffice to handle these synthetically important features. In addition the ready access to ring features that is provided by WLN is far superior to that obtainable in connection table-based systems where complex and time-consuming ring perception algorithms are needed for reactions involving ring changes

* at the time the work started, in the early 1970s, many organisations had WLN-based structure files and it thus seemed that any results might be transferred rapidly to operational systems.

The first project (Clinging and Lynch, 1973) involved creating dictionaries of simple functional group changes and then searching for the corresponding symbol changes in the WLNs of the reacting molecules. A dictionary of 50 reaction types produced analyses for about 18 per cent of the test file that was studied but the quality of the analyses was rather variable since the immediate environment of the reaction centres was not adequately described in many cases. In addition, the hyperbolic distribution of reaction types meant that a significant increase in the success rate would require a quite drastic increase in the size of the dictionary. Subsequently, Clinging and Lynch (1974) described routines to analyse reactions in which the ring system was modified. Their approach involved the identification of the individual monocycles within each ring system followed by a comparison of the two sets of monocycles so as to identify any changes that had taken place. The procedure gave successful analyses for 22 per cent of the sample file studied but was unable to process any reactions involving simultaneous acyclic transformations or changes involving more than a single monocycle.

It was concluded that although certain types of reaction could readily be encompassed by WLN-based systems, the particular approaches that had been studied could not be extended significantly in scope. Accordingly Lynch et al. (1978) developed a series of algorithms that again involved the comparison of WLN symbol strings but that did not involve the use of any sort of dictionary. Three main classes of reactions were identified, these being reactions with no apparent change in the numbers or sizes of rings, reactions with a change in the numbers or sizes of rings, and reactions involving acyclic changes or involving molecules containing benzene rings only.

The first class of reactions consisted mainly of changes in the acyclic components of cyclic molecules although there were sometimes minor changes both inside and outside the ring brackets (the WLN symbols L, T and J). The analysis consisted of comparing the ring substituents one locant position at a time, thus allowing changes at more than one substituent position. Checks were also made for certain symbol interconversions within the ring brackets to cover such elementary reaction types as reduction of ring carbonyls and the hydrogenation of unsaturated linkages. The second class of reactions was analysed in two stages. The first of these was to identify the changes in the ring system which enabled summaries of ring changes to be produced, the procedure being based upon Clinging's algorithm. The second stage of the analysis was to determine additional changes other than those occurring in the ring systems, this being carried out using the algorithm designed for the first class of reactions; it was hence possible to provide descriptions of all the parts of the molecules that had been involved in the reaction. No algorithms were developed for the third class of reaction types although it was claimed that a procedure analogous to that used for the first class could be employed; in toto, circa 70 per cent of the reactions in the file were analysed. A trial index for the first class was produced in which the sort key was the WLN symbol strings produced by the analysis.

Although satisfactory in many respects, this work suffered from several deficiencies. Firstly, the ring change algorithm was not very specific in that there was no way of connecting the ring changes with any simultaneous changes in the substituents. Secondly, in many of the reactions, the entire substituent WLN strings were given as the analysis even though large sections of them may have been unchanged i.e., the exact site of the reaction was not specifically defined. There were also problems arising from the ordering rules of the notation.

The most sophisticated use of WLN for reaction indexing is described by Lynch and Willett (1978) who applied the fragmentation and re-building techniques described above to the WLN symbol strings of the reactant and product molecules. A multi-level fragmentation procedure was developed, the levels being chosen on the basis not only of a knowledge of the reaction types in the Sheffield file but also of the way in which WLN delineates various kinds of substructural feature. The procedures were chosen so as to enable the generation of fragments representing chemically significant groups, and thus it was hoped that the routines would deal simply and effectively with common reaction types such as functional group conversions, eliminations and additions, and ring changes. The rationale for a multi-level fragmentation was to generate fragments which were as large as possible, subject to the constraint that they represented features common to both sides of the reaction equation; once these large and common features had been identified, more specific fragmentation methods could be applied to remove further common features.

In all, four levels of fragmentation were implemented together with two re-building stages in which non-common fragments were linked together to yield a reaction site. The fragmentation stages were:

* fragmentation of a WLN into the basic ring systems and substituents

* fragmentation of non-common ring systems into the component monocycles

* fragmentation of acyclic features into linear chains and WLN branching symbols such as N, X and Y

* fragmentation of linear alkyl chains into individual methylene units

Each of these fragmentation stages was followed by a comparison of the current sets of reactant and product fragments with the re-building stages involving the linking of reacting monocycles and substituents, and of reacting branching symbols and linear chains: full details are given by Willett (1978) and by Lynch and Willett (1978).

In all, acceptable analyses were produced for _circa_ 81 per cent of the sample file studied, although this figure could have been increased to _circa_ 90 per cent by additional programming. The resulting fragments and reaction sites were then used to generate a printed index of reactions.

An evaluation was made of the retrieval effectiveness of such a printed index when compared with the effectiveness obtainable from searches of the Derwent CRDS (Bawden _et al._, 1979). It was found that the two systems gave comparable levels of performance but that both had limitations with very general queries which involved some specific substructural transformation being carried out in a range of environments.

Another approach that is based, at least initially, upon WLN representations of chemical structures has been described by Osinga and Stuart. The work involves the use of a faceted classification scheme (Osinga and Stuart, 1973) for reactions which contains eight main facets, these including addition, elimination and ring change reactions. The WLNs of the reactant and product molecules are used to generate a sort of connection table in which the atoms are represented by integer descriptors rather similar to augmented atoms (Osinga and Stuart, 1974). The two sets of descriptors are then compared and the resulting analyses used as a basis for the automatic classification of the reaction; however, the method seems to have been applied only to a file of seven reactions, one of which was incorrectly processed (Osinga and Stuart, 1976).

An alternative to the direct identification of structural differences is their detection as a result of the identification of structural similarities and this was first attempted by Armitage and Lynch (1967) and Armitage et al. (1967), similarity being defined as the largest connected set of atoms and bonds common to the structures on the two sides of the reaction equation, that is the maximal common substructure (MCS). The method was based on the generation of fragments of each structure, starting with the individual atoms of each, and, by concatenation, fragments of increasing size. At each step in the process, the fragments formed from the reactant structure were compared with those from the product structure, non-common items discarded, and growth continued in the subsequent iteration only from those fragments which were common to both. The procedure terminated when the largest possible connected set of atoms and bonds had been identified. Most of the work concentrated on acyclic structures where the building blocks of the common structure were linear chains of atoms. Once these linear substructures could be grown no further, the maximal common substructure was obtained by joining the straight chains together, thus allowing the identification of branched substructures.

The procedure is intuitively appealing in that it to some extent simulates the mental processes of a chemist who, upon scanning an equation, identifies the common features as a preliminary to pinpointing the differences. However, it was found that the complexity of the programs became quite unmanageable for all but the simplest molecules since the number of chains that needed to be considered rapidly became very large, and the work was accordingly abandoned.

Interest in the use of MCS-like algorithms for reaction indexing was re-awakened by the work of Vleduts (1977) who showed that such algorithms could provide a highly accurate and precise means of specifying the unchanged portions of reacting molecules, and hence of the changed reaction sites. Moreover, the analysis could be obtained without the need for any fragmentation and re-building, thus allowing the specification not only of the reaction centre but also of its full structural environment. The major problem with such algorithms is their high computational costs since, in essence, they involve the generation of all possible reactant substructures, all possible product substructures, and the comparison of the two sets to identify the largest substructure common to the two sides of the reaction equation.

The computational problems led Lynch and Willett (1978) to develop an approximate structure matching algorithm which identified not the MCS but one or more large common substructures which, taken together, were comparable in size with or slightly smaller than the true MCS. Their algorithm was found to be two to three orders of magnitude faster for reaction indexing purposes than an exact MCS algorithm, and thus sufficiently efficient to permit the indexing of files of reactions of non-trivial size. This speed of operation was achieved by the use of a heuristic procedure in which hash codes were generated for circular substructures centred upon each of the reactant and product atoms. The assumption was made that the identification of equal hash codes upon the two sides of the equation could be taken to imply the isomorphism of the corresponding circular substructures, an isomorphism that could be detected without the need for a full atom-by-atom search. While such an assumption is likely to be quite unrealistic for general-purpose structure matching, for example in structure or substructure searching, reacting molecules generally do have extensive substructures in common, and thus the identification of equal hash codes does in fact normally signify the corresponding equivalence. In fact, detailed failure analyses suggested that intuitively sensible reaction sites were obtained for 92.6 per cent of the sample file studied.

Willett (1980) used the reaction centres identified by this algorithm to design and implement a screening system in which not only the reaction centres but also the unchanged portions of the reacting molecules were characterised by fragment screens. Queries could be processed that involved not just substructural changes but also the presence of non-reacting features, for example the reduction of an alkenyl chain whilst a vinyl group remained unchanged, while ring changes were encompassed by including the ring descriptors produced by the WLN fragmentation procedure described above. An evaluation of the search system using over a hundred queries culled from a variety of sources (Willett, 1978) resulted in a screenout of >99 per cent and a precision of _circa_ 65 per cent for those searches which retrieved any material at all from the file of _circa_ 4500 reactions used in the evaluation. This latter figure could, of course, have been increased considerably by the use of an atom-by-atom search routine as an addendum to the screening systems. These results suggest that an MCS algorithm, even the approximate one that was used, results in the identification of reaction centres that are sufficiently precise to enable chemical reaction searches to be carried out which are comparable in effectiveness to those obtainable from retrieval systems for chemical structures. Although this work was never implemented in any operational system, its results have influenced the design of several of the systems described subsequently in this volume, all of which involve (in)exact structure matching followed by a screen search on reacting and non-reacting features.

The work had been undertaken originally to provide a rapid screening mechanism for an exact MCS search. This was studied by McGregor and Willett (1981) who used the approximate reaction sites as the input to an exact MCS algorithm which was then used to identify the individual bonds that had been broken or created during the reaction. Similar work has been described by Bersohn and Mackay (1979), while further developments are described by Blower and Dana in their chapter.

CONCLUSIONS

It will be clear that the identification of appropriate mechanisms for the indexing of chemical reactions has been a protracted and difficult process, stretching as it does over many years. Indeed, even research into the relatively new automatic indexing methods has been carried out for over 20 years now. However, it would seem that relatively effective and efficient procedures *are* now available to systems designers, as will be seen in the systems and services that are described in the subsequent chapters.

REFERENCES

Armitage, J.E. and Lynch, M.F. (1967). 'Automatic detection of structural similarities among chemical compounds.' *Journal of the Chemical Society (C)*, 521-8.

Armitage, J.E., Gowe, J.E., Evans, P.N., Lynch, M.F. and McGuirk, J.A. (1967). 'Documentation of chemical reactions by computer analysis of structural changes'. *Journal of Chemical Documentation*, *7*, 209-15.

Ash, J.E., Chubb, P.A., Ward, S.E., Welford, S.M. and Willett, P. (1985). *Communication, Storage and Retrieval of Chemical Information*. Chichester: Ellis Horwood.

Bawden, D., Devon, T.K., Jackson, F.T., Wood, S.I., Lynch, M.F. and Willett, P. (1979). 'A qualitative comparison of Wiswesser Line Notation descriptors of reactions and the Derwent Chemical Reaction Documentation Service.' *Journal of Chemical Information and Computer Sciences*, *19*, 90-93.

Bersohn, M. and Esack, (1976). 'A computer representation of synthetic reactions.' *Computers and Chemistry*, *1*, 103-7.

Bersohn, M. and Mackay, K. (1979). 'Steps toward the automatic compilation of organic reactions'. *Journal of Chemical Information and Computer Sciences*, *19*, 137-41.

Clinging, R. and Lynch, M.F. (1973). 'Production of printed indexes of chemical reactions. I. Analysis of functional group interconversions.' Journal of Chemical Documentation, 13, 98-102.

Clinging, R. and Lynch, M.F. (1974). 'Production of printed indexes of chemical reactions. II. Analysis of reactions involving ring formation, cleavage and interconversion.' Journal of Chemical Documentation, 14, 69-71.

Cohen, B.J. (1982). 'User-oriented approach to a computerized reaction catalog.' Journal of Chemical Information and Computer Sciences, 22, 195-200.

Eakin, D.R. and Hyde, E. (1973). 'Evaluation of on-line techniques in a substructure search system' in Wipke, W.T., Heller, S.R., Feldmann, R.J. and Hyde, E. (eds) Computer Representation and Manipulation of Chemical Information. New York: John Wiley.

Fugmann, R., Kusemann, G. and Winter. J.H. (1979). 'The supply of information on chemical reactions in the IDC System.' Information Processing and Management, 15, 303-23.

Guthrie, R.D. (1975). 'A suggestion for the revision of mechanistic designations.' Journal of Organic Chemistry, 40, 402-7.

Harrison, J.M. and Lynch, M.F. (1970). 'Computer analysis of chemical reactions for storage and retrieval.' Journal of the Chemical Society (C), 2082-7.

Hendrickson, J.B. (1975). 'Systematic synthesis design. IV. Numerical codification of construction reactions.' Journal of the American Chemical Society, 97, 5784-5800.

Hendrickson, J.B. (1976). 'Systematic synthesis design.' Topics in Current Chemistry, 62, 49-172.

Kresze, G. (1970). 'Present-day communication in chemistry - problems and possibilities.' Angewandte Chemie International, 9, 545-50.

Lynch, M.F., Nunn, P.R. and Radcliffe, J. (1978). 'Production of printed indexes of chemical reactions using Wiswesser Line Notations.' Journal of Chemical Information and Computer Sciences, 18, 94-6.

Lynch, M.F. and Willett, P. (1978). 'The production of machine-readable descriptions of chemical reactions using Wiswesser Line Notation.' Journal of Chemical Information and Computer Sciences, 18, 149-54.

Lynch, M.F. and Willett, P. (1978). 'The automatic detection of chemical reaction sites.' Journal of Chemical Information and Computer Sciences, 18, 154-9.

McGregor, J.J. and Willett, P. (1981). 'Use of a maximal common subgraph algorithm in the automatic identification of the ostensible bond changes occurring in chemical reactions.' Journal of Chemical Information and Computer Sciences, 21, 137-40.

Meyer, E. (1975). 'Information science in relation to the chemist's needs,' in Ash, J.E. and Hyde, E. (eds) Chemical Information Systems. Chichester: Ellis Horwood.

Mishchenko, G.L. (1971). 'Information retrieval in the field of reactions of organic chemistry.' Zhurnal Vsesoyaznogo Khimicheskogo Obschestva im D.I. Mendeleyeva, 16, 55-63.

Mishchenko, G.P., Vleduts, G.E. and Shefter, A.M. (1965). 'Automatic indexing of reactions in an information retrieval system for organic chemistry.' Nauk. Tekh. Inform., 10, 13-17.

Osinga, M. and Stuart, A.A.V. (1973). 'Documentation of chemical reactions. I. A faceted classification.' Journal of Chemical Documentation, 13, 36-9.

Osinga, M. and Stuart, A.A.V. (1974). 'Documentation of chemical reactions. II. Analysis of the Wiswesser Line Notations.' Journal of Chemical Documentation, 14, 194-8.

Osinga, M. and Stuart, A.A.V. (1976). 'Documentation of chemical reactions. III. Encoding the facets.' Journal of Chemical Documentation, 16, 165-71.

Patterson, A.M. and Bunnet, J.F. (1954). 'Systematic names for substitution reactions.' Chemical Engineering News, 32, 4019.

Satchell, D.P.N. (1977). 'The classification of chemical reactions.' Naturwissenschaften, 64, 113-21.

Schier, O., Nubling, W., Steidle, W. and Valls, J. (1970). 'A system for the documentation of chemical reactions.' Angewandte Chemie International, 9, 599-604.

Schier, O. and Valls, J. (1975). 'Chemical reaction indexing' in Ash, J.E. and Hyde, E. (eds) Chemical Information Systems. Chichester: Ellis Horwood.

Valls, J. (1973). 'Reaction documentation' in Wipke, W.T., Heller, S.R., Feldmann, R.J. and Hyde, E. (eds) Computer Representation and Manipulation of Chemical Information. New York: John Wiley.

Vleduts, G.E. (1963). 'Concerning one system of classification and codification of organic reactions.' Information Storage and Retrieval, 1, 117-46.

Vleduts, G.E. (1977). Development of a Combined WLN/CTR Multilevel Approach to the Algorithmic Analysis of Chemical Reactions in View of their Automatic Indexing. London: British Library Research and Development Department.

Weygand, C. (1938). Organisch-chemische Experimentierkunst. 3 vols., Leipzig: Barth.

Weyl, T.H. (1901). Die Methoden der Organische Chemie. 3 vols., Leipzig: Thieme.

Willett, P. (1978). Computer Analysis of Chemical Reaction Information for Storage and Retrieval. Ph.D thesis, University of Sheffield.

Willett, P. (1980). 'The evaluation of an automatically indexed, machine-readable chemical reactions file.' Journal of Chemical Information and Computer Sciences, 20, 93-6.

Ziegler, J.J. (1966). 'A new information system for organic reactions.' Journal of Chemical Documentation, 6, 81-9.

Ziegler, H.J. (1979). 'Roche Integrated Reaction System (RIRS). A new documentation system for organic reactions.' Journal of Chemical Information and Computer Sciences, 19, 141-9.

2

Reaction information needs of the synthetic chemist

D. P. J. Pearson
Imperial Chemical Industries PLC, Plant Protection Division

ABSTRACT

The needs of the synthetic organic chemist for information on chemical reactions are discussed taking into account the various types of synthetic problem which are encountered. The important characteristics of a computerised reaction index are then a reflection of the needs of chemists to solve these types of problem.

This talk is based on my personal views and does not necessarily reflect the views of ICI as a whole.

Stated simply the role of the synthetic organic chemist is to make compounds. There are, however, many different environments in which synthetic problems are encountered. The approaches adopted are then governed by the type of problem being encountered. Consider the total synthesis of a natural product which usually entails the synthesis of a single, complex structure, or at most a small number of closely related materials. The synthetic route required is likely to involve a fairly large number of steps. The reactions to be used must normally be carried out with a high degree of control of stereo- and regiochemistry in a molecule which probably contains several sensitive functional groups. For instance, the molecule $PGF_{2\alpha}$ contains 5 stereocentres, 2 double bonds, one E and one Z, and its hydroxyl groups render it sensitive to a wide variety of reaction conditions.

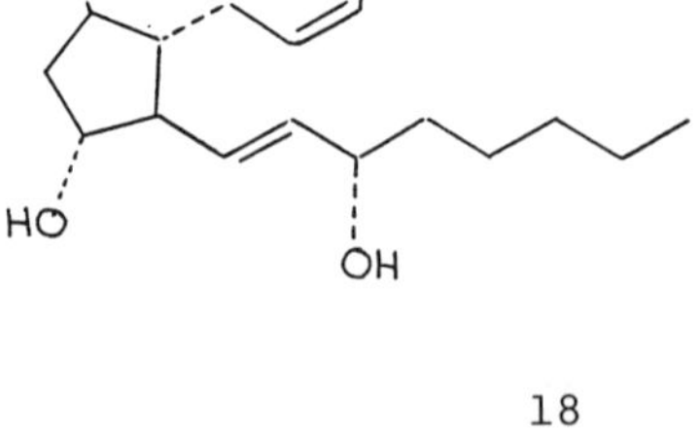

OH CO$_2$H HO OH

$PGF_{2\alpha}$

It is also to be hoped that the completed synthesis will not simply be a compilation of known reactions but that it will have an element of originality and have 'elegance'. The major exponents of this sort of chemistry are, of course, academics. However, it is becoming more common for the fine chemicals industry to become involved in the synthesis of increasingly complex molecules.

More frequently, industrial research chemists are interested in the synthesis of a range of analogues of an active compound. In a highly competitive environment such syntheses have to be carried out as efficiently as possible. Usually, many closely related structures are required. The route(s) chosen must be flexible and preferably short. This type of work is almost exclusively the province of industrial chemists.

Another type of industrial synthetic chemistry is process development which is an extremely important activity for any commercial organisation. Usually, only a single compound is required but a high degree of cost-effectiveness is required of the synthesis. A short route from cheap starting materials is needed. The reagents and solvents used need to be cheap and preferably recoverable. The safety of large industrial plants is of paramount importance so the reactions used must be as safe as possible.

The development of new chemistry must not be neglected. As previously mentioned there is a significant element of this in a total synthesis. However, it also helps commercial organisations to find novel processes to important compounds, as well as to make new structural types more readily available. This activity is important to enable organic chemistry to continue to develop as a subject. It is also necessary to capitalise on new reagents or novel intermediates and not leave them festering in the literature as mere curiosities.

These various activities as practised by organic chemists pose many different problems. The path of a synthesis from idea to execution requires the consideration of a great many factors and probably the examination of a considerable number of reactions. Perhaps this quotation from T.S. Eliot sums it up:

> 'Between the idea
> And the reality
> Between the motion
> And the act
> Falls the Shadow.'
>
> T.S. Eliot

When all is said and done the chemist must either acquire or invent a recipe if he is to achieve his synthetic goal. Perhaps in an ideal world a chemist's demands could be satisfied by a high yielding, fast reaction that is specific for the change that must be brought about. In addition, it should be easy to perform, requiring no special conditions or apparatus or exotic reagents. This is the holy grail for most chemists, and it almost goes without saying that we do not live in an ideal world.

Given these constraints which the chemist must apply to his synthetic problem, where does he get his solutions from? In order to identify useful reactions the chemist uses a large number of sources:

* personal files
* primary literature
* abstracting journals such as Chemical Abstracts, Current Chemical Reactions, etc.
* collections such as those of Theilheimer or Fieser and Fieser
* reviews and books.

Most chemists use some form of personal files, even if it is only their memories. More normally it is a card index, a notebook, or in a few cases a microcomputer system. They also use a standard range of published sources, often incorporating the data from these sources into their own index. These sources in my experience fall into the classes listed. Time precludes a detailed discussion of their merits and demerits: suffice it to say that no single source satisfies all the needs of all chemists. There are indeed a plethora of useful sources, most of them overlapping to some degree which makes any attempt at a full coverage impossible. This makes chemists very prone to favouritism!

Let us consider the chemist's 'card file' for the moment. A study of this type of data storage should illustrate a chemist's minimum requirements of a reaction index - if it does not do what his present 'system' does it will be abandoned fairly quickly. The advantages of the personal file are:

* It is selective and much trivial chemistry is omitted. It is to be hoped that only those reactions viewed by the end user as useful have been included.
* Being mainly compiled from the primary literature it is usually up-to-date.
* It is always available on the chemist's desk.
* Manually entering reactions into the system has an imprinting effect on the memory making recall easier.
* Chemists like hard copy.

The disadvantages are:

* The file cannot be comprehensive and at each moment in time tends to reflect the individual's current interests. This can degrade its value at a later date.

* Very few individuals manage to maintain any cross-referencing system, e.g. using punched cards.

* Structural indexing is restricted to functional group/compound type name fields only.

* It takes time to maintain and leads to a lot of duplication of effort.

This means that although such systems are useful, their utility is limited and search methods are often memory dependent. This leads to the conclusion that a computerised reaction index with the correct characteristics should be an extremely useful tool for the synthetic organic chemist, and though I doubt it would ever completely replace current methods, its success in replacing them depends on many factors not least of which would be credibility.

What problems must a reaction index solve?

1. The preparation of a structure from no particular starting material, generalised as:

$? \longrightarrow A$

This is equivalent to consulting a review on the preparation of A. For instance, how can I prepare 4-substituted isoquinolines, 1,5-dicarbonyl compounds, or alpha-keto-esters? It should be possible within this category to specify bonds which must be made or those preserved during the reaction. This would enable the chemist to select only those reactions which suit his overall strategy, and would be equivalent to looking for functional group selectivity.

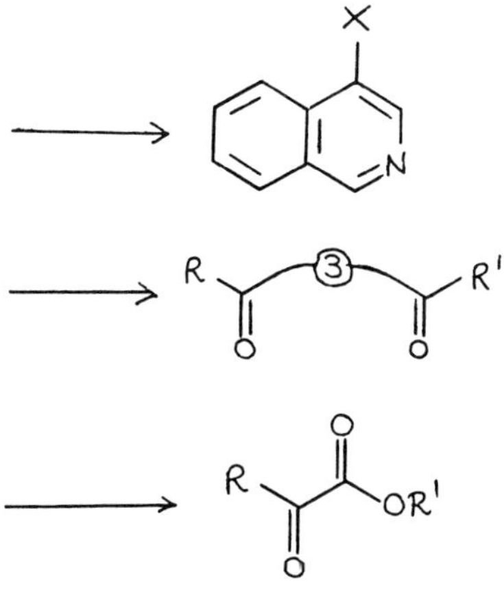

X

N

R ③ R'

O O

O

R OR'

O

2. The preparation of one structural type from another, i.e.

A $\longrightarrow$ B

This type of question is usually much more specific but can be complicated by the constraints which the chemist wishes to apply. Perhaps it is desired to synthesise an aldehyde from a nitrile in less than four steps. It is also essential that some facility for searching out stereo-selective or stereospecific reactions be provided, not only for relative stereochemistry but also for absolute stereochemistry. The system should reliably handle geometrical isomers, i.e. the specific reduction of cyclohexanones. It is essential that searching for functional group selectivity be possible, i.e. how to reduce a ketone in the presence of an aldehyde.

RCN $\longrightarrow$ RCHO

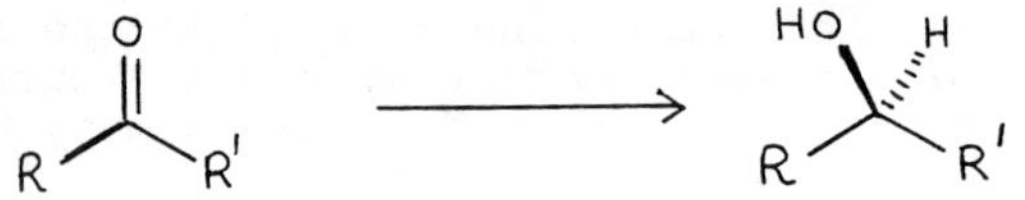

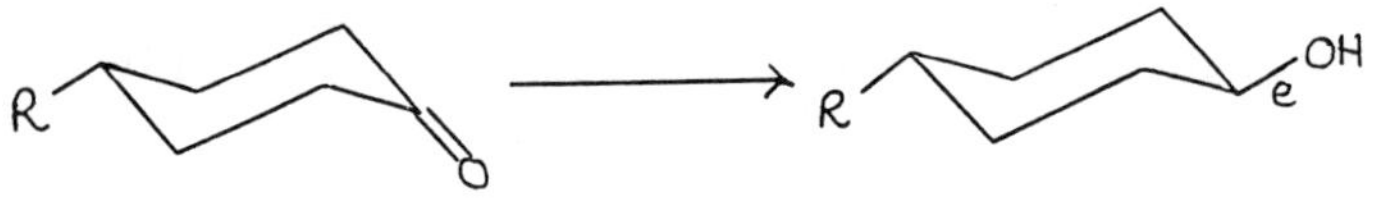

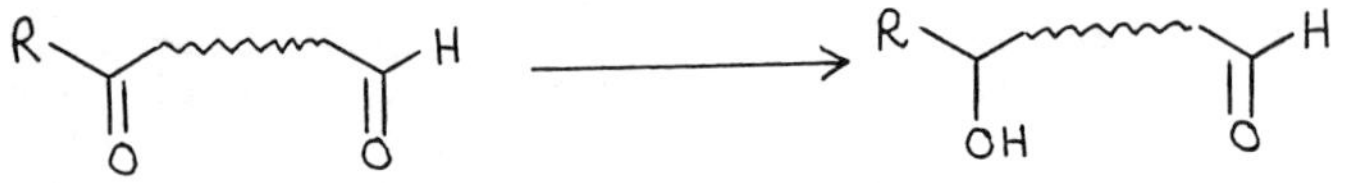

3. What is the product of a reaction between two molecules? i.e.

A + B $\longrightarrow$?

This option is extremely useful if interference of functional groups is anticipated for a given reaction or if more than one product is possible. If the searcher has not already found a reaction of A and B which leads to his desired product this option is useful for checking whether this reaction is known to lead to an incorrect product. The reaction of 2-aminopyridine with beta-ketoesters can give either of the two products shown.

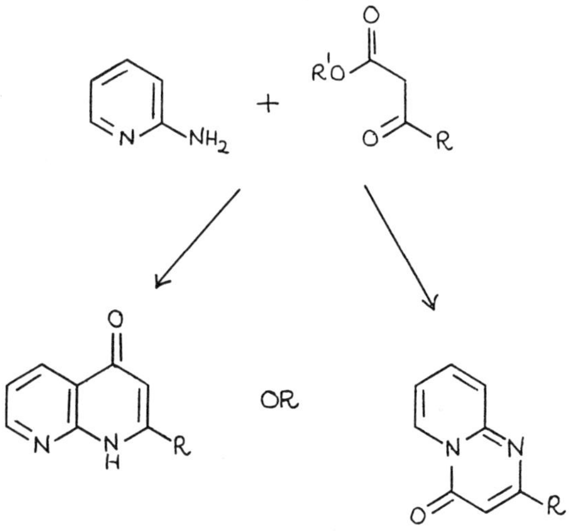

4. What uses has reagent A been put to?

$$? \xrightarrow{A} ?$$

This is perhaps best regarded as a subset of 3, for example what synthetic uses can chloro-iminocarbamates be put to? The search for close precedents should help to define possible products for a potentially useful intermediate. However, the solutions to other problems may depend on the policy adopted on the indexing of reagents. Reagent oriented searches are definitely required as information on non-acidic hydrolyses or non-metallic oxidising agents, for example, is sometimes required. The use of keywords may be the most appropriate solution to this problem.

There are several other outstanding problems which need to be tackled by a reaction index in order to satisfy chemist needs.

1. Multistep reactions are often encountered in the literature. Because an isolated abstractor considers a single sequence of importance does not mean to say that the individual steps in the sequence should not be of interest to other chemists. In the sequence converting a nitrile to an aldehyde, shown below, two intermediates are found. A chemist may also wish to know how to make nitrilium salts or imines from nitriles and if an indexer has assumed too much this information will be lost. There are, of course, cases where to index each step of a reaction sequence individually would be ridiculous. The indexing of synthetic sequences also carries a consequence concerning the accuracy of any reaction centres of which more later.

$$RCN + Et_3\overset{\oplus}{O}\ \overset{\ominus}{BF_4} \longrightarrow RC{\equiv}\overset{\oplus}{N}Et\ \overset{\ominus}{BF_4}$$

$$\xrightarrow{Et_3SiH} RCH{=}NEt \xrightarrow{H_2O} RCHO$$

83-99%

2. The most useful reactions are generally those with the most examples, some papers publishing as many as fifty or so analogous preparations. The choice of the correct examples from a set will determine the quality of a database. For instance, the reduction of anilines using Ti(II) can be done in the presence of a number of other groups, and it would be sensible to select the example which would normally be most sensitive to conditions used for nitro reduction. The other possible solution to this selection problem is a generic representation of reactions. Without this type of solution it should be said that the chemist will judge a reaction index by the quality of coverage the database has: hence the importance which must be placed on reaction selection, which should normally be done by a practising chemist.

$$X{-}C_6H_4{-}NO_2 \xrightarrow{[Ti(II)]} X{-}C_6H_4{-}NH_2$$

X = Cl, CN, $-C(=O)OCH_2CH{=}CH_2$

3. The <u>reaction centre</u> is a concept central to most reaction indexes, enabling more specific searches to be made. This too presents problems: does an indexer rely on a mechanistic interpretation of a reaction or does he simply identify bonds which have changed during the reaction? Personally, I favour a relatively rigorous approach based on the participating atoms rather than bonds; this should result in greater flexibility. Otherwise, some reactions become difficult cases, for example the carboxylation of benzene derivatives really only needs a single atom reaction centre on the starting material. If we consider the multi-step case again, the tertiary-butoxylation of bromothiophene is clear enough but when the alkyl group is removed it is the <u>alkyl</u> group that is now involved in the reaction centre. So to represent the overall reaction as shown carries with it an element of ambiguity and using the intermediate from the first step as a template for the second would obviously be incorrect.

1. MeLi
2. CO_2

1. Mg
2. $PhCO_2O$+

$H^{\oplus}$

2 Steps

4. The handling of tautomers presents the chemist with another barrier to the use of a reaction index since he will tire of having to carry out several searches to achieve a 'complete' search. This would amount to two passes for 2-pyridone, but many more possibilities exist for more complex molecules such as those shown below. Even a partial solution to this problem would be an important feature of any system.

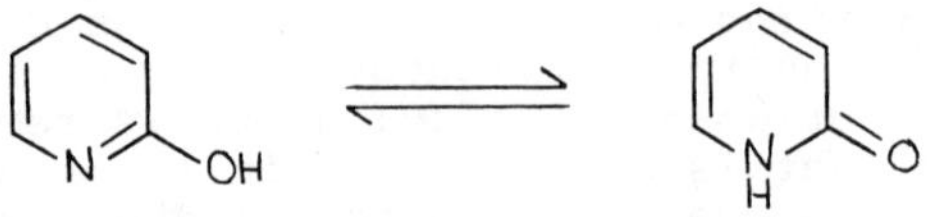

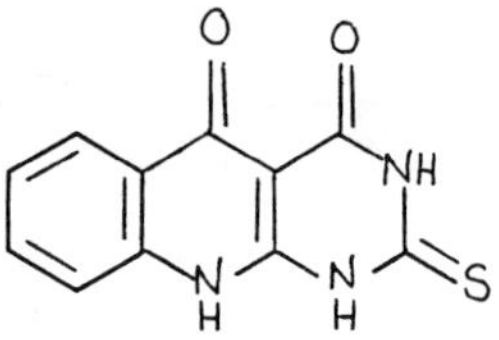

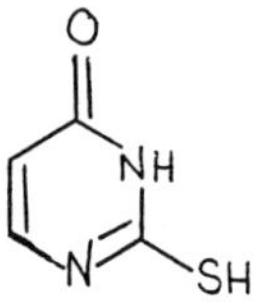

5. What should such a system look like to ensure making a favourable impression on the potential user? I think that we can take for granted that it should use graphics and be able to carry out structure and sub-structure searches. The system should enable chemists to pose questions in the way in which they are already used to thinking. This requires easy graphical input and the easy selection of any structural or text qualifiers. The output format should be flexible to allow for the different needs of various types of user. However, as a default it should have a concise reaction diagram, report the yield of the product, show any reagents used, display a reference and show comments on the reaction such as the examples reported, spread of yields etc.

6. A reaction index should be easy to use, fast to search and the terminals should not go 'dead' while the computer is searching.

To sum up, the system requirements are:

* Good quality general coverage by database of ca. 50000+ reactions.
* Fast search times.
* Cosmetically chemist compatible.
* Easy to use.
* Up-to-date.
* Handles stereochemistry, tautomers and preferably generic structures.
* Allows searches for selective reactions.
* Gives results with high signal-noise.
* Handles data at same speed as structures.
* Little or no 'down' time so that the system is always available.

3 Printed tools for reaction searching

I. Sinclair
Pfizer Central Research

The average research chemist devotes a large proportion of his time to literature searching. Three main areas where literature searching is necessary have been identified:

1. The generation of alternative routes by retrosynthetic analysis. Searching for more economic routes, or routes where the use of toxic reagents is avoided. Alternative routes may also be sought to avoid problems of patent infringements. Computerised synthesis programs such as LHASA, SECS and EROS could be used for this type of search.

2. The evaluation of the feasibility of routes in the synthetic direction. The successfulness of individual steps may be investigated, also whether the specified reaction conditions will lead to the desired products and whether there are any interfering functionalities. Programs such as Professor Jorgensen's CAMEO may be of use here.

3. The retrieval of synthetic information from the published literature. This is important in order to keep abreast of current research, and also in looking for support for current ideas by finding good analogies.

It is in this third area that chemists are most likely to be found literature searching, and that this paper will concentrate on. Special emphasis will be given to selected examples of printed tools available, rather than attempting a comprehensive coverage.

It seems that in general the minimum amount of training is given to chemists in literature searching with less than 35% of US colleges and universities offering formal courses. Many chemists experience only 2-3 hours of training in seven years, this mainly being concerned with compound searching. PhD students tend to be introduced to searching by looking first at previous PhD theses, and then current journals, with this usually leading to personal card indexes. Card indexes can be criticised as being poorly organised, not very systematic and certainly not comprehensive, but they do form invaluable sources for reaction searching. Advanced text books are also good information sources e.g. "Advanced Organic Chemistry" by March contains 600 reactions and over 10,000 references. These beginnings are important as they give good leads into the primary and review literature.

What can be discerned about approaches to literature searching is that methods tend to be highly dependent on the individual chemist, who will favour those sources which have given good results in the past.

Examples of how information on different compounds or reactions can be found, serve to outline the advantages and disadvantages of existing printed tools.

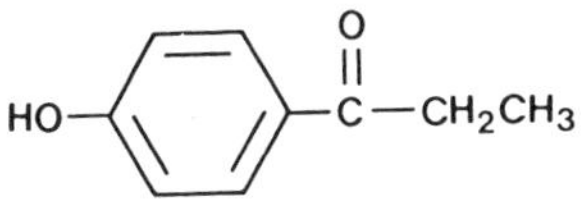

p-Hydroxypropiophenone

4'-Hydroxypropiophenone

1-(4-Hydroxyphenyl)-1-propanone

p-Propionylphenol

Ethyl p-hydroxyphenyl ketone

Figure 1.

Here information is needed on a relatively well known compound. Using the Merck Cross Reference Index six separate references to the primary literature are found, each involving some variation on the Fries Rearrangement. In this case a further alternative name, Paroxypropione, is found.

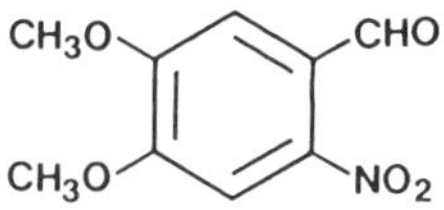

6-Nitroveratraldehyde

4,5-Dimethoxy-2-nitrobenzaldehyde

Figure 2.

Again this is a relatively well known compound. In this case the Dictionary of Organic Compounds is useful, giving a reference to the nitration of veratraldehyde.

Phenylalanine **P-00921**

α-Aminobenzenepropanoic acid, 9CI. *2-Amino-3-phenylpropanoic acid*. *β-Phenylalanine*

COOH
H_2N—C—H (*S*)-*form*
CH_2Ph

$C_9H_{11}NO_2$ M 165

(***R***)-***form*** [673-06-3]

D-form

Leaflets (H_2O). Mp 283-4° dec., 196° dec. $[\alpha]_D^{25}$ +34.5° (c, 1 in H_2O), +4.47° (c, 1 in 5M HCl).

▷AY7533000.

N-*Formyl:* Plates (H_2O). Mp 167°. $[\alpha]_D^{26}$ −72.7°.

N-*Ac:* Mp 170-1°. $[\alpha]_D$ −49.3°.

N-*Benzoyl:* Needles (H_2O). Mp 146-8°. $[\alpha]_D^{20}$ −17.1° (KOH aq.).

N-*Et; B,HCl:* Mp 154-6°.

N-*Benzenesulphonyl:* Mp 133°. $[\alpha]_D$ +6.7°.

N-p-*Toluenesulphonyl:* Cryst. (EtOH aq.). Mp 164-5°. $[\alpha]_D^{29}$ +2.4° (Me_2CO).

N-*Chloroacetyl:* Mp 125°. $[\alpha]_D$ −50.8° (c, 2 in EtOH).

Me ester; B,HCl: $[\alpha]_D^{25}$ +3.9° (c, 2.7 in H_2O).

(***S***)-***form*** [63-91-2]

L-form

Widely distributed in proteins. Needles (H_2O). Mp 283-4° (rapid heat). $[\alpha]_D^{25}$ −34.5° (c, 1 in H_2O), −4.5° (c, 1 in 5M HCl). *N*-Protected derivs. useful in peptide synth. have been listed alphabetically elsewhere.

▷AY7535000.

Et ester; B,HCl: Cryst. (EtOH/Et_2O). Mp 154-6°. $[\alpha]_D^{20}$ −7.6° (c, 3 in H_2O).

Me ester; B,HCl: Mp 159-61°. $[\alpha]_D^{25}$ −4.6° (c, 5 in H_2O).

Benzyl ester; B,HCl: Mp 203°.

N-*Formyl:* Plates (H_2O). Mp 167°. $[\alpha]_D^{20}$ +75.2° (EtOH).

N-*Ac:* Mp 170-71°. $[\alpha]_D$ +48.2°.

N-*Benzoyl:* Mp 142-3°. $[\alpha]_D$ +19.8°.

N-*Benzenesulphonyl:* Mp 133°. $[\alpha]_D$ −7°.

N-*2,4-Dinitrophenyl:* Cryst. (MeOH aq.). Mp 189°.

Figure 3.

This entry shows a further use of this Dictionary to check quickly for reported derivatives. In this case, what N-protected phenylalanines have been reported. The lack of references to actual reactions can, however, be a problem.

$HO{-}CH_2{-}C{\equiv}C{-}CO_2CH_3$

Methyl γ-hydroxytetrolate

Methyl 4-hydroxy-2-butynoate

Figure 4.

This is an example where the chemist would probably refer to the Chemical Abstracts Formula Index.

$C_5H_6O_2$ — CHEMICAL ABSTRACTS——10th COLL. — 1298F

ethenyl ester, polymer with butyl 2-propenoate [61488-60-6], 86: P 18326w
ethenyl ester, polymer with 2-chloro-1,3-= butadiene [67517-90-2], 89: P 111966b
ethenyl ester, polymer with ethenyl acetate [27380-80-9], 88: P 201090v; 91: P 202236w; 92: P 85986y
ethenyl ester, polymer with 1-ethenyl-2-= pyrrolidinone and methyl 2-methyl-2-= propenoate [67101-26-2], 89: P 111740y
ethenyl ester, polymer with isooctyl 2-propenoate and (tetrahydro-2-furanyl)methyl 2-propenoate [78733-28-5], 95: P 86339p
ethenyl ester, polymer with methyl 2-propenoate [27102-10-9], 86: P 172168c
ethenyl ester, polymer with 2-(phosphonoox= y)ethyl 2-methyl-2-propenoate [67021-20-9], 89: P 25333m
ethenyl ester, polymer with 1*H*-pyrrole-2,5-dione [74079-09-7], 93: P 29792r
2-Propen-1-one, 1-oxiranyl- [56209-32-6]. For general derivs. see *Chemical Substance Index*
Propylidene, 1-acetyl-2-oxo- [53121-07-6], 93: 7172m
2-Propynoic acid
ethyl ester [623-47-2]. See *Chemical Substance Index*
2-Propyn-1-ol
acetate [627-09-8], 86: 15966u, 16270z, 43612u, 71634w, 120530v, P 157829a; 88: 22073v, 121285x; 89: 108229h; 90: 71367u, 87332q, 103491y, 125487d; 91: 51857w, 174495j; 92: P 146801h; 93: 26351k, 45482v; 94: 155997v, 185276b; 95: 6392n, 80608d, 169077c
2*H*-Pyran-2-one, 3,4-dihydro- [26638-97-1], 87:

(*Z*)- [70143-04-3], 90: 187155c
——, 4-oxo-
methyl ester [7327-99-3], 91: P 19992a, 210486t; 92: 22097p; 93: P 8335d; 95: 97502n
methyl ester, (*E*)- [5837-72-9], 86: P 43248e, P 72921z, 89237d, 140256p; 91: 11318q, 123304p, 192773c; 92: 128237z; 93: 72002v; 95: 17994m, 97523v
methyl ester, (*Z*)- [57314-32-6], 89: 146373h
2-Butyne-1,4-diol
monoformate [69989-70-4]. For general derivs. see *Chemical Substance Index*
→ 2-Butynoic acid, 4-hydroxy-
methyl ester [31555-05-2], 90: 39057g; 91: 38979t; 92: P 75874w; 94: 31036m
3-Butynoic acid, 2-hydroxy-
methyl ester [70861-76-6], 91: P 85293z; 95: 76048d
——, 2-hydroxy-2-methyl-
(±)- [71748-56-6], 91: 170607z
(*S*)- [33675-93-3], 95: 128249a
Carbonic acid
diethenyl ester [7570-02-7], 93: P 7667b
methyl 2-propynyl ester [61764-71-4], 86: 71634w
1,3-Cyclopentanedione, 4-hydroxy- [62966-15-8], 87: 22481f
2-Cyclopenten-1-one, 2,3-dihydroxy- [80-72-8], 86: 113997z; 88: P 161434b, 186291r; 89: 6478w; 90: 163980a; 91: 119469d, 188646w; 92: 179066a, 198705z; 93: 112417t, P 248172a; 94: P 90386y; 95: P 1896v
Cyclopropaneacetic acid, α-oxo- [13885-13-7], 89: P 5991w
2-Cyclopropen-1-one, 2,3-dimethoxy-

——, 4-methoxy- [65990-45-6], 88: 136387f
——, 5-methoxy- [74118-39-1], 93: 46267r
2,4-Oxetanedione, 3,3-dimethyl- [22922-58-3], 89: 129034f; 93: 7095p; 95: P 6539r
Oxiranecarboxylic acid
ethenyl ester, polymer with ethene and ethenyl acetate [68183-22-2], 89: P 216470p
2,4-Pentadienoic acid, 2-hydroxy- [50480-68-7], 90: 50459k
2,3,4-Pentanetrione [921-11-9], 86: 105604d; 89: 42040f
2-Pentenal, 5-hydroxy-4-oxo- [73982-80-6], 93: 26673a
2-Pentenoic acid, 4-oxo- [4743-82-2], 86: 42854n, P 89190h; 88: 5795q; 90: 22505d; 91: P 56360p; 92: 14907n; 93: 150231k; 95: 61129g, 203505c, 209287e
(*E*)- [2833-28-5], 90: 134866d, 168012n; 94: P 103331d
3-Pentenoic acid, 2-oxo-
(*E*)- [68982-84-3], 90: 50459k
lithium salt, (*E*)- [68982-86-5], 90: 50459k
4-Pentenoic acid, 2-oxo- [20406-62-6], 89: 17661b, 190979j; 90: 50459k, 115048x; 91: 102884n, 119351j, 186760e
DL-*threo*-2-Pentulose, 1,4:3,5-dianhydro- [75870-07-4], 94: 16019q
DL-*threo*-3-Pentulose, 1,2:4,5-dianhydro- [63941-87-7], 87: 102559c
erythro-3-Pentulose, 1,2:4,5-dianhydro- [63976-11-4], 87: 102559c
2-Pentynoic acid, 4-hydroxy-
(±)- [70404-01-2], 91: 5129n
(+)- [76293-75-9], 94: 65068p
3-Pentynoic acid, 2-hydroxy-

Figure 5.

But, as can be seen from Figure 5, problems do arise in finding the correct indexing name.

$HO{-}CH_2{-}C{\equiv}CH$ + [3,4-dihydro-2H-pyran] → [tetrahydropyranyl]–O–$CH_2{-}C{\equiv}CH$

↓

$HO{-}CH_2{-}C{\equiv}C{-}CO_2CH_3$ ← [tetrahydropyranyl]–O–$CH_2{-}C{\equiv}C{-}CO_2CH_3$

ORG. SYN., 60, 81 (1981)

Figure 6.

Knowing this however, references to a synthesis can be found. This can be updated using Science Citation Index, or alternatively the Registry Number could be used in an Online Search (CAS). Both methods lead to an entry in Organic Synthesis (1981) as shown in Figure 6.

Note that this does suggest that Organic Synthesis should have been used in the first place. Knowing which tools are available, and which to use for a particular case is obviously very important. Thus Beilstein and the Chemical Abstracts Chemical Substances Index shown in Figures 7-9 both allow a certain degree of browsing.

Hydroxycarbonsäuren $C_4H_4O_3$

Hydroxy-but-2-insäure $C_4H_4O_3$ = HO-CH_2-C≡C-CO-OH (H 390; E III 713; dort auch als Hydroxymethyl-propiolsäure und als 4-Hydroxy-tetrolsäure bezeichnet).

B. Beim Behandeln von Prop-2-in-1-ol mit Äthylmagnesiumbromid in Benzol und Behandeln des Reaktionsprodukts mit Kohlendioxid (*Henbest et al.*, Soc. **1950 3646, 3649;** *Smith, Jones*, Canad. J. Chem. **37** [1959] 2092; vgl. H 390). Beim Behandeln einer Lösung von 2-Prop-2-inyloxy-tetrahydro-pyran in Tetrahydrofuran mit Äthylmagnesiumbromid in Äther, mit Kohlendioxid und anschliessend mit wss. Schwefelsäure (*He. et al.*). Beim Erhitzen von Acetoxy-but-2-insäure mit Wasser (*Dupont et al.*, Bl. **1954 816,** 819).

Krystalle (aus Bzl.); F: 115–116° (*Du. et al.*), 115° (*He. et al.*).

Acetoxy-but-2-insäure $C_6H_6O_4$ = CH_3-CO-O-CH_2-C≡C-CO-OH.

B. Beim Behandeln einer Lösung von 4-Acetoxy-but-2-in-1-ol in Aceton mit Chrom(VI)-oxid und wss. Schwefelsäure (*Dupont et al.*, Bl. **1954** 816, 819).

Krystalle (aus Bzl.); F: 66°.

Hydroxy-but-2-insäure-methylester $C_5H_6O_3$ = HO-CH_2-C≡C-CO-O-CH_3.

B. Beim Behandeln einer Lösung von 2-Prop-2-inyloxy-tetrahydro-pyran in Benzol mit Äthylmagnesiumbromid in Äther, mit Kohlendioxid und anschliessend mit Methanol und wenig Schwefelsäure (*Henbest et al.*, Soc. **1950** 3646, 3650).

$Kp_{0,5}$: 73°. n_D^{20}: 1,4712.

Methoxy-but-2-insäure-äthylester $C_7H_{10}O_3$ = CH_3-O-CH_2-C≡C-CO-O-C_2H_5.

B. Beim Behandeln von 3-Methoxy-propin mit Äthylmagnesiumbromid in Äther und anschliessend mit Kohlensäure-diäthylester (*Gillespie, Price*, J. org. Chem. **22** [1957] 780, 783).

Kp_3: 63–64°. n_D^{25}: 1,4397. IR-Banden im Bereich von 3 μ bis 14 μ: *Gi., Pr.*

Figure 7.

pyrolysis of, **95:** 187044u
ethyl ester *[59938-67-9]*
prepn. and addn. reaction with anthracene, **86:** 4624e
prepn. and reaction of, with anthracene, **95:** 203028z
2-Propynoic-*3-^{14}C* **acid**
——, **3-(1,3-benzodioxol-5-yl)-** *[62848-92-4]*
prepn. of, and condensation with oxalyl chloride, **86:** 189625x
1-Propyn-1-ol
acetate, platinum complex *[78325-53-8]*, NMR of, MO calcns. in relation to, **95:** 41714r
2-Propyn-1-ol (*propargyl alcohol*) *[107-19-7]*
86: P 189199m; **88:** R 6239y, P 6321u
acetalization of acetaldehyde and butyraldehyde by, **93:** 94759z
acetalization of paraformaldehyde with, **89:** 107985q
acetylation of, with acetic acid, **90:** 151253u
in acid cleaner for locomotives, and treatment of wastewater therefrom, **87:** P 73076k
acidity, hydrogen bonding, and reactivity of, IR in relation to, **87:** 133667r
acidity of, in DMSO, dynamic NMR study of, **88:** 151613v
addn. of
with diisopropyl ketone, **91:** 4826a
with mercaptans, **88:** P 50273d
addn. of chloride to, in presence of acetonitrile or DMF, mechanism of electrochem., **90:** 203192v
addn. reaction of
with α-acetylenic ketones, **88:** 22296v
with alkenylzinc bromides, **88:** 104610g

Figure 8.

presence of nickel chelates, **88:** 170531h
alkyl ethers, prepn. and hydrolysis of, **88:** 190556y
alkynylation by, of transition metal halides, copper halide-catalyzed, **87:** 135840r
alkynylation of Et iodobenzoate with, **87:** 22693b
alkynylation of halopyridazines by, **95:** 24957e
aminomethylation of, **87:** 22307d
bactericidal activity of, freezing in relation to, **94:** 150986f
benzoylation of, with amine catalysts, kinetics of, **88:** 88885t
benzylation of, with benzyl bromide, **88:** 189999p
bromination and etherification of, **94:** P 174372d
bromination of
86: P 30278q; **87:** P 22386d; **94:** P 65098y
with phosphorus tribromide, propargyl bromide by, **93:** 94737r
Cadiot-Chodkiewicz reaction of, with bromo(methylenedioxyphenyl)acetylene, **89:** 75299a
carbon-13 NMR of
93: 149216c, 149217d
effect of cross-polarization time on, **89:** 107200e
isotope shifts in, **93:** 203465y
carbonylation of, **93:** 45925s
→ carboxylation of, by di-Me carbonate, **92:** P 75874w
catalytic hydrogenation of, in presence of phosphorus oxy acid, **95:** P 168536q
chemisorption of, on palladium, hydrogenation catalysis in relation to, **89:** 49463e
chlorination of
with benzyltriphenoxyphosphonium chloride, **94:** 174187x
with cupric chloride, solvent effect on, **87:** 67794b
mechanism and stereochem. of, **87:** 38753g
with thionyl chloride, propargyl chloride by, **93:** 94738s
O-Chloromethylation of, by aldehydes and hydrogen chloride, **88:** P 6579j
Chodkiewicz coupling of, with (methylbenzyl)= bromoundecynamide, **95:** P 52501v
CH wagging vibration in, **86:** 71637z
clathrate hydration of, **94:** 14954s
codimerization of, with allyl halides in presence of palladium complex catalysts, **90:** 54406p
competitive hydrogenation with, on nickel catalyst, **90:** 167725k
condensation of
with butylchloronitrobenzamide, **90:** P 6118z

Figure 9.

For example, in the case of the CACSI, a search can be based on the starting material of the reaction in Figure 6 (Propargyl Alcohol or 2-Propyn-1-ol). This leads to a synthesis using dimethyl carbonate.

Figure 10 demonstrates four alternative routes by which a novel compound may be synthesised. The type of reaction involved determines which tool should be used.

For route 1, an allylic oxidation, examples of useful tools are:

Theilheimer
Fieser and Fieser
Monographs on Oxidation (Augustine et al)
Houben-Weyl
Modern Synthetic Reactions
Organic Functional Group Preparations
Harrison and Harrison Compendium
Synthetic Reagents (for SeO2)
Advances in Organic Chemistry
Chemical Abstracts General Subject Index

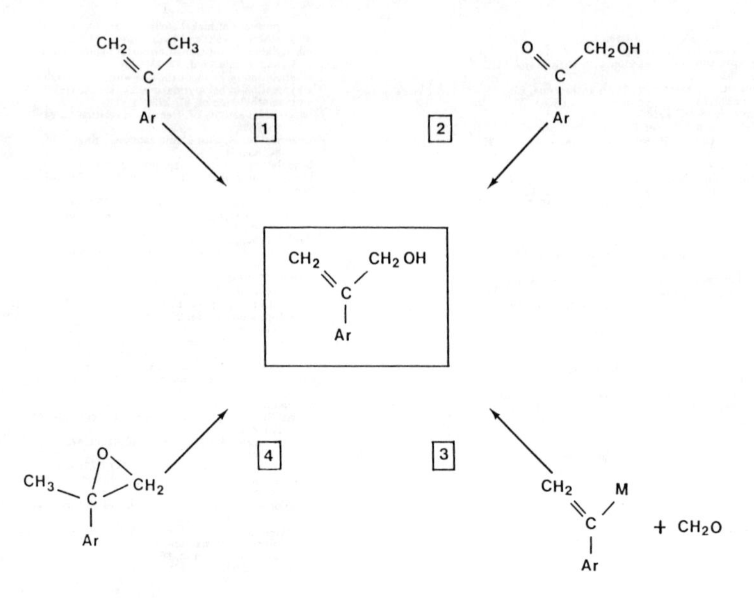

Figure 10.

Route 2 is a Wittig type reaction. It is important to look for reviews. Useful sources would be:

Organic Reactions
Synthesis
Angewandte Chemie
Annual Reports on Organic Synthesis
General Synthetic Methods

Route 3 involves a vinyl organometallic. Good tools may be:

Annual Reports on Organic Synthesis
Carbanions in Organic Synthesis (Stowell)
Comprehensive Carbanion Chemistry (Durst)
Carbanions in Chemistry (Bates and Ogle)

Route 4 could be present in:

Chemistry of the Ether Linkage (and supplement)
Weissberger Heterocyclic Series
Organic Reactions (recently reviewed)

The best method for simple aryls is not easily found and Figure 11 shows a further useful method. Organometallic chemistry is a particularly difficult area to cover. Comprehensive Organometallic Chemistry is very good, particularly chapters by Magnus (Org Si), Trost (Pd) and Pearson (Organo Iron). References to this reaction could be found under Propargyl Alcohol in the CA Substances Index, but would the chemist think to look here? Another useful tool is an ICI book, New Pathways for Organic Synthesis. This deals with applications of transition metals.

$$HC{\equiv}C{-}CH_2OH \; + \; ArMgX \xrightarrow{Cu\,(I)} CH_2{=}C(Ar){-}CH_2OH$$

B.Jousseaume, J.Organomet.Chem., 168, 1 (1979)

Figure 11.

More Complex Problems

Examples of far more complex problems can be given. How can a chemist find information on Diastereoselective Aldol Reaction of alpha-Fluoro Ketone Enolates, or Enantiospecific Syntheses from carbohydrate precursors or amino acids/ terpenes.

Special topics are now arising, such as phase transfer catalysis and photo/electro-chemical procedures, each of which has an extensive literature of its own.

Conclusions

It can be seen that the printed tools available for reaction searching are numerous. They do, however, present certain disadvantages. These search systems are generally non-structural, unfriendly and not conducive to browsing. Language barriers are also present. The English chemist would tend to avoid a tool such as Beilstein in favour of something perhaps less comprehensive, but produced in English. It is clear that although current awareness is extremely important to the Organic Chemist, present manual systems do present problems in accessing this vital information. Furthermore, the chemist's approach to the literature tends to be highly individualistic, depending on training, past experience and personal likes.

In applying computers to the task of reaction searching it is unlikely that any one system can answer all the needs of chemists. Certain important features must, however, be included. Firstly, the system must be graphically based, structure entry being achieved using a light pen or mouse. Secondly, the system should have potential for browsing through a database, something which is highly impractical across large volumes of printed tools; unconstrained searching is invaluable in overcoming preconceived ideas which can obstruct efficient searching.

4 The Chemical Reactions Documentation Service - pastures old and new

A. Finch
Derwent Publications Limited

INTRODUCTION

William Theilheimer's renowned yearbooks Synthetic Methods of Organic Chemistry (SMOC)(1) came to light some forty years ago at a time when information on chemical reactions was becoming increasingly difficult to locate in the ever-expanding literature. Thus was developed a unique system for gathering and presenting synthetic methods in the framework of an annual series which has stood the test of time. Ensuring continuity of SMOC is now the responsibility of Derwent Publications Ltd., which ten years ago introduced the Journal of Synthetic Methods (JSM)(2) from which SMOC is now derived. It is the latter series up to Volume 30 (1974) and the JSM, from 1975 to date, which provide the database for the Chemical Reactions Documentation Service (CRDS), the essential features of which are described in this paper.

THE CRDS DATABASE

The CRDS database currently totals more than 60,000 reactions, covering such aspects as:

novel functional group and ring transformations of a general nature,

new reagents and synthetic techniques, and

new syntheses of key ring systems.

Each year 3000 new synthetic methods are added to the file. These data correspond to the abstracts (Figure 1) appearing in the current issues of the JSM, for which reactions are selected by careful scrutiny and analysis of the worldwide chemical journal and patents literature. In addition to abstracts on novel methods, the JSM also reports on approximately 3000 minor modifications of known reactions, which appear biannually in the form of Supplementary Reference Indexes. Applications of known reactions, however, are not within the scope of the CRDS, nor are specialised fields of chemistry, such as the syntheses of esoteric heterocycles, which are relatively simple to access by conventional means (3).

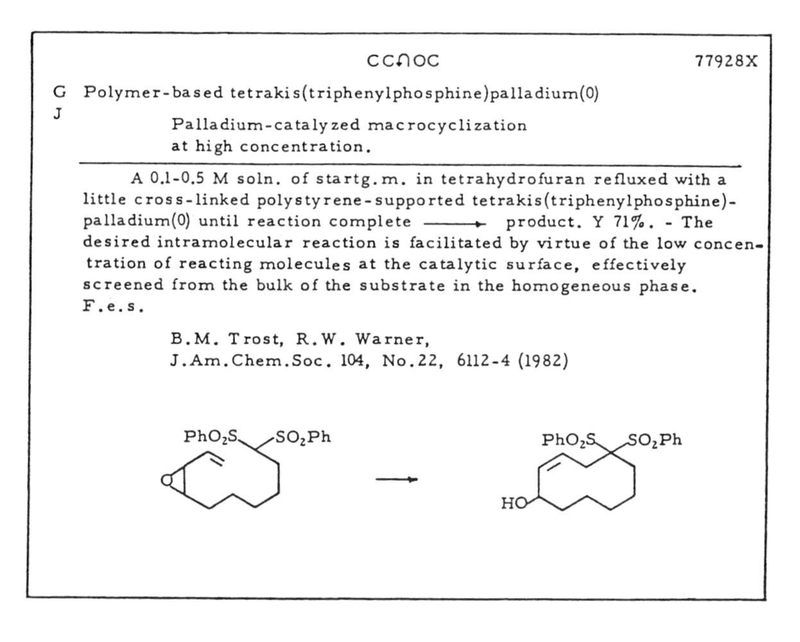

CCΩOC 77928X

G Polymer-based tetrakis(triphenylphosphine)palladium(0)

J

Palladium-catalyzed macrocyclization at high concentration.

A 0.1-0.5 M soln. of startg.m. in tetrahydrofuran refluxed with a little cross-linked polystyrene-supported tetrakis(triphenylphosphine)-palladium(0) until reaction complete ⟶ product. Y 71%. - The desired intramolecular reaction is facilitated by virtue of the low concentration of reacting molecules at the catalytic surface, effectively screened from the bulk of the substrate in the homogeneous phase. F.e.s.

B.M. Trost, R.W. Warner, J.Am.Chem.Soc. 104, No.22, 6112-4 (1982)

Figure 1. Sample abstract in the Journal of Synthetic Methods.

RETRIEVAL SYSTEMS

For access to the computerised information in the CRDS database, two methods of retrieval are currently available: the Coding and the Keywording. An outline of these systems is given below.

The Coding System

This, the earlier of the two retrieval systems, was developed in the late sixties by a consortium of European pharmaceutical companies, and originally designed for in-house retrieval of the reactions in SMOC. Originally based on the 80-column punch card, it is a formal system by which the key reaction parameters - starting materials, products, reagents, bond breakages and formations, and reaction conditions - are input and retrievable as three-digit codes.

As an illustration (Figure 2), the strategy for the reduction of azido to amino groups is a combination of six codes.

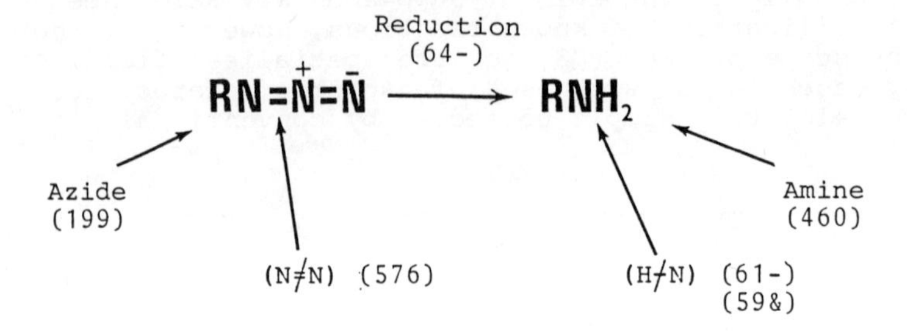

Strategy:

```
SS 1 /C?
USER:
199 AND 576 AND 61- AND 460 AND 59& AND 64-
```

Figure 2. Example of a search strategy using the Coding.

On interrogation of the complete CRDS database (>60,000 reactions) from 1942-1984, thirty-six answers were obtained with a relevance of 85%. The printout of the eight most recent references (Figure 3) shows for each record the corresponding JSM (or SMOC) abstract number, the title (as assigned for the abstract by Derwent or Theilheimer), and the citation. The Theilheimer reaction symbol notation (featured in the abstracts above the title) appears in the printout after the title. This is often valuable in scanning the printout for relevance, and may be included in the search strategy itself, as illustrated later under Keywording (Figure 8)(4).

The Keywording

This, the second method of retrieval, was introduced in 1978 and was specifically designed for online usage (5).

There is a controlled list of some 1000 keywords for starting materials and products, and around 700 terms for generic reagent classes, reaction conditions and reaction types. These are supported by 20 single-letter Reaction Classes (Thematic Groups) for retrieval of broad reaction types, e.g. R for reduction, H for heterocyclic ring closure, and Z for protective group chemistry.

Keywords for starting materials and products (Figure 4) are of two main types: functional group keywords, e.g. ALCOHOL, AMINE, ISOCYANATE, and ring keywords, e.g. OXAZOLE,

```
SS 1 /C?
USER:
199 AND 576 AND 61- AND 460 AND 59& AND 64-

SS 1 PSTG (36)

SS 2 /C?
USER:
PRT 8

PROG:

-1-
AN - 77509Z RY
TI - /AMINES FROM AZIDES UNDER MILD CONDITIONS . SELECTIVE REDUCTION/
     HN-X-N / /
CI - CHEM.LETTERS 1984, NO.10, 1733-6.

-2-
AN - 77760Y R
TI - /AZETIDINES FROM 2-AZETIDINONES/ HC-X-O / /
CI - J.AM.CHEM.SOC. 105, NO.20, 6339-42 (1983).

-3-
AN - 76252Y R
TI - /AMINES FROM AZIDES . TRANSFER HYDROGENATION/ HN-X-N / /
CI - TETRAHEDRON LETTERS 24, NO.15, 1609-10 (1983).

-4-
AN - 77508X R
TI - /AMINES FROM AZIDES/ HN-X-N / /
CI - CHEM.IND. 1982, NO.18, 720.

-5-
AN - 77507X R
TI - /AMINES FROM AZIDES . PHASE TRANSFER CATALYSIS/ HN-X-N / /
CI - J.ORG.CHEM. 47, NO.22, 4327-9 (1982).

-6-
AN - 76101W R
TI - /CYCLIC AMIDINES FROM IMINOHALIDES VIA IMINOAZIDES .
     4-AMINOPYRIMIDINES/ NC-X-G, HN-X-N / /
CI - J.ORG.CHEM. 46, NO.7, 1413-23 (1981).

-7-
AN - 75503W R
TI - /AMINES FROM AZIDES UNDER NEUTRAL CONDITIONS . SELECTIVE
     REDUCTION . AR. AMINES/ HN-X-N / /
CI - J.CHEM.RES.(S) 1981, NO.1, 17.

-8-
AN - 77753V R
TI - /AMINES FROM AZIDES . HETEROCYCLIC PRIM . AMINES . ALSO FROM
     HYDRAZINES/ HN-X-N / /
CI - SYNTHESIS 1980, NO.10, 830-31.
```

Figure 3. Printout of the first 8 references corresponding to the search strategy given in Figure 2.

```
AMINE
AMINE,AR
.
.
.
C-ACID
C-ACYLHALIDE
C-AMIDE
C-ANHYDRIDE
.
.
.
DIOXOLE-1,2
DIOXOLE-1,3
DIOXY
DIOXY-1,1
DIOXY-1,2
.
.
.
```

```
SI-ACID
SI-AMINE
SI-AMIDE
SI-ETHER
SI-ETHER+
SI-HALIDE
SI-HYDRIDE
SI-THIOACID
SI-THIOETHER
SI-THIOETHER+
SILANE
SILANE+
SILANE+
SILICIC
SILICON
SILYL
SILYL,TRIMETHYL
.
.
.
```

Figure 4. Sections of the Keyword list.

CYCLOBUTANE. Where possible, nomenclature is such as to permit retrieval of related keywords in a single operation by right-hand truncation. Thus, both aliphatic amines (AMINE) and aromatic amines (AMINE,AR) may be retrieved together by searching for the fragment AMIN in the following manner

ALL AMIN:

Similarly, all silicon functions (except DISILANE) may be retrieved with

ASS SI:

and all carboxylic acid derivatives by

ALL C-:

Included in the keyword list are a number of so-called 'structural' and 'adjectival' keywords. With the former, one can locate substructure fragments of a general nature, such as -O-C-N-, -O-C-C-O-, -N-C-C(=Y)-, and thereby perform broad searches without specifying the precise nature of the functional group(s), e.g. the structural keyword DIOXY-1,2 (for the fragment -O-C-C-O-) retrieves both glycols (HO-C-C-OH) and their simple derivatives (RO-C-C-OR, TsO-C-C-OTs, etc.) in one operation. Adjectival keywords, on the other hand, were designed primarily for refining purposes. Thus, while it is possible to retrieve all 1,4-dioxins and hydrogenated derivatives with the keyword DIOXIN-1,4, one can simply add an adjectival keyword (UNSATD, DIHYDRO, etc.) for greater specificity, thus

DIOXIN-1,4 is used for 1,4-dioxins (unsaturated or hydrogenated)

DIOXIN-1,4 AND UNSATD is used for unsaturated 1,4-dioxins

DIOXIN-1,4 AND DIHYDRO is used for dihydro-1,4-dioxins

DIOXIN-1,4 AND TETRAHYDRO is used for 1,4-dioxanes

Specific reagents are retrieved by using standardised Theilheimer nomenclature, e.g. MERCURIC-ACETATE, 1,8-DIAZABICYCLO[5.4.0]UNDEC-7-ENE, whereas for reagent classes there is a controlled list of reagent keywords, e.g. HG-2 for all mercury(II) compounds and N-BASE,DIAZABICYCLIC for all diazabicyclic compounds as base.

In order to define the role of keywords, the following search qualifiers are available:

SM for starting materials

PR for products

RT for reagents and reaction terms, and

TG for reaction classes (thematic groups)

Thus, for the reduction of aliphatic azides to amines:

$$RN_3 \longrightarrow RNH_2 \qquad (1)$$

R=aliphatic

the search strategy would be

AZIDE-C/CM AND R/TG AND AMINE/PR

Inputting the latter in the CRDS keywording database of >45,000 reactions (1966-1984) gave a response of 36 postings with a relevance of 100%. In the full printout of the first hit, abstract 77509Z (Figure 5), one can see that starting material keywords appear within oblique strokes and product keywords within asterisks. Of particular note are keywords enclosed by both asterisks and brackets, e.g. *(KETONE)*, which illustrates the indexing of an unaffected group. For retrieval of the latter, the following strategy is used:

'KETONE)' /PR

Thus, for the selective reduction of aldehydes in the presence of keto groups:

$$RCO....CHO \longrightarrow RCO....CH_2OH \qquad (2)$$

the following strategy could be used:

```
SS 3 /C?
USER:
PRT AN TI CI KW 1

PROG:

-1-
AN  - 77509Z RY
TI  - /AMINES FROM AZIDES UNDER MILD CONDITIONS . SELECTIVE REDUCTION/
      HN-X-N / /
CI  - CHEM.LETTERS 1984, NO.10, 1733-6.
KW  - /AZIDE-C/ GIVES *AMINE,AR* *AMINE* INORG-REDN TE
      (SODIUM-HYDROGEN-TELLURIDE) SOLV=4 TEMP=4 *(KETONE)* *(ETHYLENE)*
      *(ACETYLENE)* *(C-ACID)* *(C-AMIDE)* *(C-ESTER)* *(NITRILE)*
      *(HALIDE,AR)* *(HALIDE)* *(SULFONE)*
```

Figure 5. Sample Keywording record in full.

```
SS 1 /C?
USER:
ALDEHYDE/SM AND ALCOHOL/PR AND 'KETONE)' /PR AND R/TG

PROG:
SS 1 PSTG (27)

SS 2 /C?
USER:
PRT 27

PROG:
:
-4-
AN  - 76757Z RY
TI  - /PRIM . ALCOHOLS FROM ALDEHYDES WITH RETENTION OF KETO GROUPS/
      HC-A-OC / /
CI  - TETRAHEDRON LETTERS 25, NO.28, 2985-6 (1984).

-5-
AN  - 76511Z RY
TI  - /ALCOHOLS FROM OXO COMPDS . PREFERENTIAL REDUCTION/ HC-A-OC / /
CI  - TETRAHEDRON LETTERS 24, NO.48, 5367-70 (1983).

-6-
AN  - 75515Z RY
TI  - /PRIM . ALCOHOLS FROM ALDEHYDES/ HC-A-OC / /
CI  - J.ORG.CHEM. 49, NO.1, 163-6 (1984).

-7-
AN  - 75171Z RNXY
TI  - /ALPHA,BETA-DIHYDROXYKETONES FROM ALPHA-KETOALDEHYDES AND
      ALDEHYDES VIA TIN(II) ENEDIOLATES . CROSS-ALDOL CONDENSATION/
      CC-A-OC / /
CI  - CHEM.LETTERS 1983, NO.12, 1825-8.

-8-
AN  - 77260Y RY
TI  - /PRIM . ALCOHOLS FROM ALDEHYDES WITH RETENTION OF KETO GROUPS/
      HC-A-OC / /
CI  - TETRAHEDRON LETTERS 24, NO.40, 4287-90 (1983).
:
```

Figure 6. Printout of the first 8 references from the Keywording search for the selective reduction of aldehydes in the presence of keto groups (equation 2).

ALDEHYDE/SM AND ALCOHOL/PR AND R/TG AND 'KETONE)' /PR

The response to this search in the keywording database was 27 postings with a relevance of 74% (Figure 6).

For ring closures of a general nature, it is normally sufficient to search for the functional groups involved at the reaction centre in combination with the thematic group for ring closure (H for heterocyclic, J for carbocyclic ring closure). Thus, halogenolactonization,

(3)

is readily retrieved as follows:

ETHYLENE/SM AND C-ACID/SM AND HALIDE/PR AND LACTONE/PR AND H/TG

One could also have included among the products the structural keyword OXYHALIDE-1,2 for the fragment -O-C-C-Hal, but in this instance the four keywords indicated with H/TG give a manageable response (13 postings; 90% relevance) so that refinement with additional keywords is not necessary (Figure 7).

For simple reactions - often the most important for the practising chemist - the choice of keywords may be so limited that refinement of the search, i.e. elimination of noise, may be impossible by using controlled terms. In such instances, the Theilheimer reaction symbol may be included in the search strategy, as illustrated in Figure 8 for the conversion of alcohols to nitriles. Here, the initial keyword response was 29, but subsequent inclusion of the reaction symbol (CC-X-O) - which was identified by inspecting the first few answers during the course of the search - dramatically reduced the noise level.

Trends in chemistry over a particular period may also be discerned by keyword searches. This is illustrated in Figure 9, a graphical representation of palladium-catalyzed synthesis of ethylene derivatives from O-allyl compounds (equation 4) over the period 1975-1984.

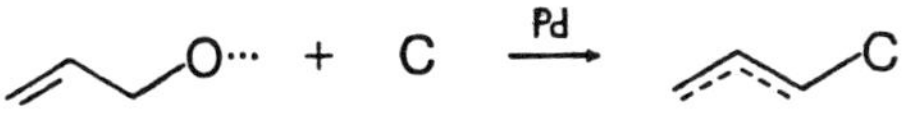

(4)

```
SS 1 /C?
USER:
ETHYLENE/SM AND C-ACID/SM AND LACTONE/PR AND HALIDE/PR AND H/TG

PROG:
SS 1 PSTG (13)

SS 2 /C?
USER:
PRT 13

PROG:

-1-
AN   - 77368Z H
TI   - /2-ALPHA-HALOGENO-O-HETEROCYCLICS FROM ETHYLENEALCOHOLS/ GC-X-H / /
CI   - J.CHEM.SOC. CHEM.COMMUN. 1984, NO.16, 1070-1.

-2-
AN   - 75117Y HO
TI   - /5-ALKYL-2(5H)-FURANONES FROM ALPHA-ALKYLTHIO-GAMMA,DELTA-
       ETHYLENECARBOXYLIC ACIDS VIA IODOLACTONIZATION/ GC-X-H / /
CI   - BULL.CHEM.SOC.JAPAN 55, NO.12, 3935-6 (1982).

-3-
AN   - 76875X HX
TI   - /BROMOLACTONIZATION/ GC-X-H / /
CI   - ARCH.PHARMACAL.RES. 4, NO.2, 133-5, 137-9 (1981) (ENG); CHEM.ABSTR.
       97, 38807, 38808 (1982).

-4-
AN   - 77874W YHX
TI   - /DELTA-IODO-GAMMA-LACTONES FROM GAMMA,DELTA- ETHYLENECARBOXYLIC
       ACIDS . REGIO- AND STEREO-SPECIFIC IODOLACTONIZATION . ALSO
       CONVERSION TO THREO-BETA-HYDROXY- GAMMA,DELTA-OXIDOCARBOXYLIC ACID
       ESTERS/ GC-X-H / /
CI   - TETRAHEDRON LETTERS 22, NO.46, 4611-14 (1981).

-5-
AN   - 77820W HLG
TI   - /GAMMA-LACTONES FROM CYCLOPROPANECARBOXYLIC ACID ESTERS VIA
       GAMMA-IODOCARBOXYLIC ACIDS . THIS IS PART OF A 3-STEP SYNTHESIS
       OF GAMMA-LACTONES FROM ETHYLENE DERIVS/ OC-E-G, GC-X-C / /
CI   - TETRAHEDRON LETTERS 22, NO.49, 4891-94 (1981).

-6-
AN   - 77364W H
TI   - /CHLOROLACTONIZATION . DELTA-CHLORO-GAMMA-LACTONES . EPSILON-
       CHLORO-DELTA-LACTONES/ GC-X-H / /
CI   - J.ORG.CHEM. 46, NO.17, 3552-54 (1981).
:
```

Figure 7. Printout of the first 6 references from the Keywording search for halogenolactonization (equation 3).

```
SS 1 /C?
USER:
ALCOHOL/SM AND NITRILE/PR

PROG:
SS 1 PSTG (29)

SS 2 /C?
USER:
PRT TI 5

PROG:

  -1-
X TI  - /REGIOSPECIFIC SYNTHESIS OF ALPHA-AMINOKETONES FROM AMINES AND 2
        DIFFERENT ALDEHYDE MOLECULES VIA ALPHA-AMINONITRILES AND
        ALPHA-AMINO-BETA-HYDROXYNITRILES/ OC-E-C, CC-X-O, CC-A-OC / /

  -2-
√ TI  - /NITRILES FROM ALCOHOLS VIA FORMIC ACID ESTERS . BENZYL NITRILES/
        CC-X-O / /

  -3-
X TI  - /REPLACEMENT OF SULFINYL BY CYANO GROUPS . CYANOMETHYL ETHERS .
        THIS IS PART OF A MULTISTEP SYNTHESIS OF CYANOMETHYL ETHERS FROM
        ALCOHOLS/ CC-X-S / /

  -4-
√ TI  - /NITRILES FROM ALCOHOLS UNDER MILD CONDITIONS . WITH INVERSION OF
        CONFIGURATION/ CC-X-O / /

  -5-
X TI  - /OXIDATIVE CYANO GROUP MIGRATION . CYANOCARBOXYLIC ACIDS FROM CYCLI
        CYANOHYDRINS/ OC-A-CC / /

SS 2 /C?
USER:
1 AND CC-X-O

PROG:
SS 2 PSTG (11)

SS 3 /C?
USER:
PRT TI 5

PROG:

  -1-
X TI  - /REGIOSPECIFIC SYNTHESIS OF ALPHA-AMINOKETONES FROM AMINES AND 2
        DIFFERENT ALDEHYDE MOLECULES VIA ALPHA-AMINONITRILES AND
        ALPHA-AMINO-BETA-HYDROXYNITRILES/ OC-E-C, CC-X-O, CC-A-OC / /

  -2-
√ TI  - /NITRILES FROM ALCOHOLS VIA FORMIC ACID ESTERS . BENZYL NITRILES/
        CC-X-O / /

  -3-
√ TI  - /NITRILES FROM ALCOHOLS UNDER MILD CONDITIONS . WITH INVERSION OF
        CONFIGURATION/ CC-X-O / /

  -4-
√ TI  - /NITRILES FROM PRIM . ALCOHOL . SYNTHESIS WITH ADDITION OF 1
        C-ATOM/ CC-X-O / /
  .
  .
  .
```

Figure 8. A section of the printout from the Keywording search for the conversion of alcohols to nitriles, illustrating the value of the Theilheimer reaction symbol (CC-X-O) in the reduction of noise level.

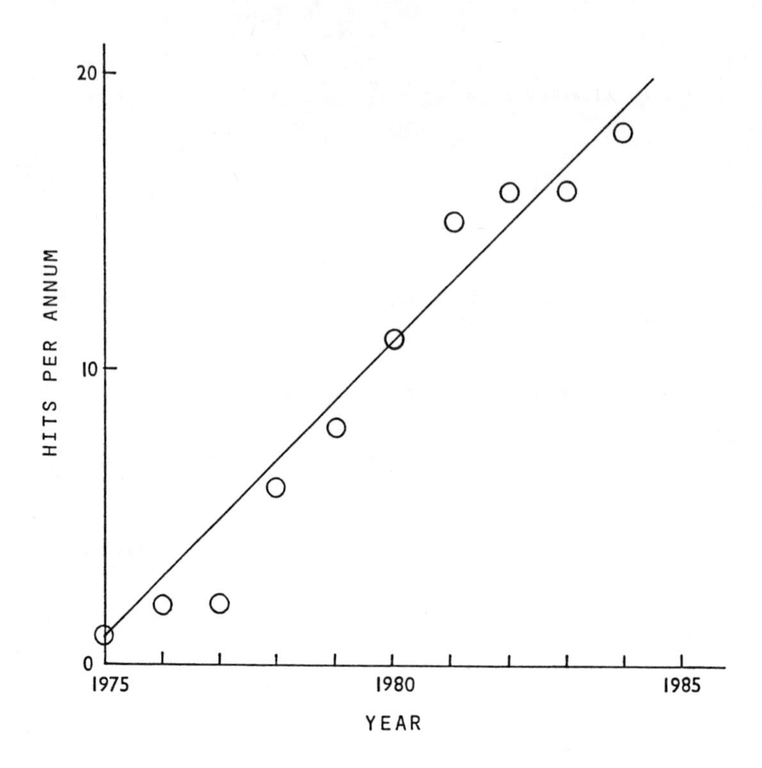

Figure 9. Graphical representation of the Keywording search for palladium-catalized syntheses with O-allyl derivatives (equation 4) covering the period 1975-1984 (30,000 reactions).

For this search the following strategy was used for each of the ten years concerned:

ALLYL-O/SM AND ALL ETHYLENE:/PR AND PD/RT

In order to facilitate retrieval by using the keywording system, a Thesaurus (Figure 10) is available and will shortly be updated to include a formula index as an alternative means of locating the keywording of functional groups. This will be an adaptation of Theilheimer's 'Formula Index of Complex Functional Groups', which provides quick access to the Theilheimer and JSM nomenclature as used in the Subject Indexes. In the Keywording Thesaurus, this index will appear in the same format but with structural formulae and the corresponding keywords for each entry in place of nomenclature (Figure 11). As an extension, the structural formulae themselves (e.g. $C(SR)SO_2R$) and functional group notation (e.g. O_2S_2C) could be readily input online together with the keywording in order to provide an independent means of accessing functional group conversions in a more

nitroaldehydes,o-	use	NITRO-C ALDEHYDE AZACARBONYL-O
nitroamines	see	nitramines
nitrocarboxylic derivs.,alpha-	use	NITRO-C AZACARBONYL-2 and appropriate keywords for the acid function (s. Pt.1, p.131)
nitrocarboxylic derivs.,beta-	use	NITRO-C AZACARBONYL-3 and appropriate keywords for the acid function (s. Pt.1, p.131)
nitrocarboxylic derivs.,o-	use	NITRO,AR AZACARBONYL-O and appropriate keywords for the acid function (s. Pt.1, p.131)
nitro compds.,aliphatic	use	NITRO-C
-,simple preparations and reactions	see	Pt.1, p.85
-as reagent	use	ALL NITRO:/RT
nitro compds.,ar.	use	NITRO,AR
nitro compds.,aci-	see	nitronates
nitro compds.,other	sa.	di- nitramines
nitroenamines,2- (>N-C=C-NO2)	use	NITRO-C ENE-1,2-DIHETERO ENAMINE
nitroethers	see	alkoxynitro compds.
nitroethylene derivs.,1,1- (C=C-NO2)	use	NITRO-C ETHYLENE-N
nitroethylene derivs.,2- (C=C-C-NO2)	use	NITRO-C ETHYLENE ALLYL-N
NITROGEN	see	Pt.1, p.127
nitrogen dioxide	use	('NITROGEN-DIOXIDE)' /RT OR 'DINITROGEN-TETROXIDE)' /RT)

Figure 10. Section from the Keywording Thesaurus.

⋮

OS_2C

Acyldisulfides	RCOSSR	DISULFIDE+	CS-ACID
Dithiolcarbonic acid esters	RS-CO-SR	THIOCARBONATE	DITHIO
1-Organothiosulfines	C(SR)=S=O	SULFINE	CS-ESTER+
1-Sulfinylthioethers	RSO-C-SR	THIOCARBONATE	DITHIO
Xanthates	RS-CS-OR		

⋮

Figure 11. Section of the Formula Index of Complex Functional Groups.

convenient manner.

GRAPHICAL RETRIEVAL

The graphics software currently being developed for reaction databases will greatly simplify retrieval of chemical reactions. Already, the reactions in Volumes 1-35 of SMOC are being converted to a graphics-readable format for in-house retrieval by the REACCS software of Molecular Design Ltd. (6), and this same company have a licence from Derwent to make available the data in the JSM in the same manner. A corresponding online graphics file is also under study at Derwent, and this would naturally enhance the CRDS considerably.

While graphics software has a clear advantage over keywording and coding systems in respect of user-friendliness, the same need not necessarily be true for overall efficiency, and some interesting comparisons between the old and the new methods are eagerly awaited when a sizeable CRDS graphics file is available. The philosophy that one can retrieve all reactions by inputting structures and reaction centres alone seems a dangerous one, and there is little doubt - at least for the SMOC/JSM data - that some natural language will be necessary for a certain proportion of reactions. As a simple illustration, it will be virtually impossible to retrieve protective group chemistry by graphics alone since, for example, input of a hydroxyl group will hardly provide all methods of hydroxyl group protection without an intolerable degree of noise. C-a-Alkylation as a concept could also be practically irretrievable, as could many simple functional group conversions where the number of parameters available is small.

Having the facility of specifying the exact stereochemistry of a molecule or substructure may seem at first sight a considerable advantage to graphics users, but in the field of chemical reactions it is not always the molecular configuration that is important; it is rather the dynamic aspect that a reaction proceeds with stereoselectivity. This is highlighted in the field of asymmetric synthesis, where a clear distinction must be made between, for example, the many stereospecific aldol-type condensations (equations 5 and 6) and the few asymmetric aldol-type condensations (equation 7), yielding only one, say (SR)-erythro, of the four possible enantiomers (7).

It is difficult to imagine how such a distinction could be made simply by using a structural input alone. What is even more significant is that research in the field of asymmetric synthesis is directed more towards the steric constraints that are used to induce asymmetry, e.g. a chiral X group in equation (7), rather than the formation of a specific optically active compound. Yet the choice of the X group could be quite arbitrary, and a simple modification of this ligand might easily induce formation of, say, the (RS-erythro-enantiomer. This principle is exemplified in a recent publication (8), where the author used a cyclic α, β-unsaturated acetal based on (S,S)-(-)-N,N,N',N'-tetramethyltartaric acid diamide for the preparation of chiral β-substituted ketones, e.g. (S)-3-methylcyclohexanone; yet it was clearly pointed out, though not illustrated, that had the (R,R)-diamide been used instead the antipodal ketones, e.g. (R)-3-methylcyclohexanone, would have resulted. The fact that asymmetry is induced is clearly far more important than the fact that a specific enantiomer is formed.

S,R erythro R,S (5)

S,S threo R,R (6)

S,R-erythro Asymmetric (7)

Me_2NOC $CONMe_2$ (S,S) Me_3Al (S) (8)

Me_2NOC $CONMe_2$ (R,R) Me_3Al (R) (9)

Herein, also, one can perhaps appreciate the subtle distinction between a chemical reaction and a synthetic method, the former presenting a somewhat static aspect of the latter which may often embody a crucial dynamic factor undefinable by structural parameters.

CONCLUSIONS

We have almost reached the stage where the synthetic chemist demands rapid and efficient means of accessing chemical reactions from comprehensive, regularly updated databases. Unfortunately, no system today will give all of the answers all of the time, and no two reaction databases are alike. Derwent's CRDS provides two options for computer-based retrieval, and a graphics equivalent will shortly be available. Each of these three systems has its merits and limitations, and knowing which to choose can come only with experience. The one aspect, however, which they have in common is the database itself - and in the final analysis it is the quality and continuity of this that matters most.

REFERENCES AND NOTES

1. Theilheimer, W. 'Synthetic Methods of Organic Chemistry" Karger: Basel.

2. The 'Journal of Synthetic Methods' is published by Derwent Publications Ltd., London, as part of the 'Chemical Reactions Documentation Service'.

3. For further details on the 'Journal of Synthetic Methods' see Finch, A.F., <u>J. Chem. Inf. Comput. Sci</u>. in press.

4. For details of the Coding system see Schier, O. et al., <u>Angew. Chem</u>., <u>Int. Ed. Engl</u>., 1970, 9, 599.

5. Both Coding and Keywording are currently accessible via System Development Corporation (SDC), 2500 Colorado Avenue, Santa Monica, California 90406, USA.

6. Molecular Design Limited, 2132 Farallon Drive, San Leandro, California 94577, USA.

7. In the context of the aldol condensation, the qualification 'stereospecific' is used for <u>erythro</u>- or <u>threo</u>-specific condensations where a racemic product is obtained; the qualification 'asymmetric' is used for enantioselective condensations.

8. Fukutani, Y. et al., <u>Tetrahedron Letters</u>, 1984, 25, 5911-2.

5 User experience of retrieving chemical reaction information from publicly available online services

P. T. Bysouth and J. Hardwick
Glaxo Group Research

INTRODUCTION

Use of chemical reactions to achieve synthetic targets is a fundamental daily routine of the laboratory chemist.

The earliest known recipes were on Egyptian papyri. The first chemistry text-book, Alchemia, written by Andreas Libavinus, appeared in 1597. Today the chemist is faced with an enormous literature into which he must regularly delve for information needed to safely and economically repeat known substance conversions or to attempt the successful synthesis of novel structures.

Whether use is made of online files or printed secondary sources, there is an eventual need to go beyond merely knowing that a substance exists or indeed that it can be synthesised (at least hypothetically) from a certain precursor, to knowing in detail HOW?

The chemist ultimately seeks knowledge of preparative conditions, solvents, reagents, isolation, purification, yields and physical data. His information goal is experimental detail as traditionally found in the primary chemical journals. Thus we need to ask whether online sources of reaction information are firstly, satisfying this need and secondly, more effective than the traditional printed secondary publications. It is important to realise that in the latter case the age of a reference does not necessarily detract from its value to the laboratory chemist.

Our information scientists currently handle in a year between 50 and 60 major reaction queries (excluding single compound look-ups). They use a wide variety of sources but in the last twelve months only incurred costs of about £1250 accessing specific reaction files. That level of expenditure in fact represents only a tiny percentage of our total online budget.

For a major pharmaceutical company employing 250 research and development organic chemists the above figures equate to just one major reaction query per chemist every five years. This ridiculously low level of activity I feel reflects the poor quality of service we can offer our chemists given present online resources. They are forced to rely on laboriously monitoring and gathering information from current awareness services and from their own reading of the primary chemical journals.

TYPES OF REACTION QUERY MOST FREQUENTLY ENCOUNTERED

We commonly receive five types of query that involve reaction searching.

1. The preparation or reaction of a specific compound.

   * By any method

   ? ⟶ Compound A

   e.g. Preparation of $ClCH_2C(Ph)(OH)CH_2Cl$

   Compound B ⟶ ?

   Search tools available: CAS ONLINE-Registry File and CA File.

   * From a specific compound

   Compound C ⟶ Compound D

   e.g. $ClCH_2C(=O)CH_2Cl \longrightarrow ClCH_2C(Ph)(OH)CH_2Cl$

   Search tools available: CAS ONLINE-Registry File and CA File.

2. The use of a specific reagent and/or specific conditions.

   e.g. Use of $NaCNBH_3$ with aldehydes

   Search tools available: CAS ONLINE-Registry File and CA File, CRDS, Reagents for Organic Synthesis Vols. 1-10 (Fieser & Fieser)

3. The preparation or transformation of a specific generic structure.

   Substructure E ⟶ ?

? $\longrightarrow$ Substructure F

Substructure G $\longrightarrow$ Substructure H

e.g.

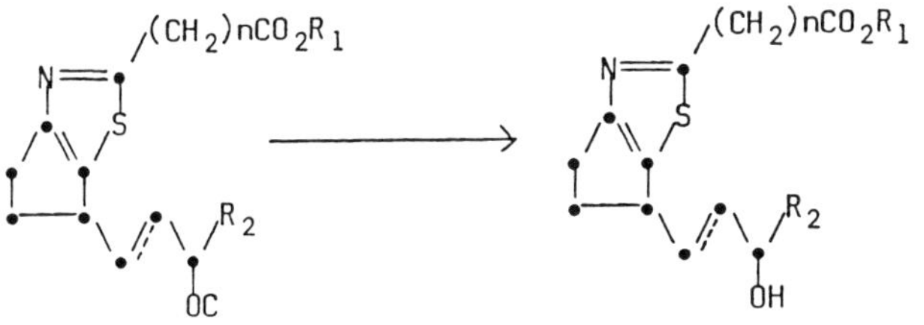

$(CH_2)nCO_2R_1$ ... $\longrightarrow$ ... $(CH_2)nCO_2R_1$

where R_1 = H or C

R_2 = anything (including H)

n = 1-10

Further substitution is allowed.

Search tools available: CAS ONLINE-Registry File and CA File.

4. Functional group transformations.

Functional group I $\longrightarrow$?

? $\longrightarrow$ Functional Group J

Functional Group K $\longrightarrow$ Functional Group L

e.g. $-CO_2R_1 \longrightarrow -CONR_2R_3$

Search tools available: CRDS, CAS ONLINE-CA File, Compendium of Organic Synthetic Methods Vols. I-VI (Harrison and Harrison)

These transformations might be required in the presence of other functional groups which must remain unaffected by the reaction conditions.

e.g.

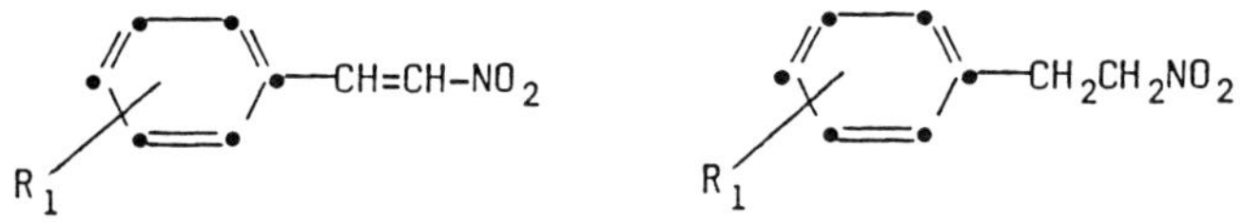

R_1–C₆H₄–CH=CH–NO_2 $\longrightarrow$ R_1–C₆H₄–$CH_2CH_2NO_2$

where R_1 = anything (including H)

Search tools available: CRDS, CAS ONLINE-CA File.

5. The use of a specified named reaction.

   e.g. References to the McFadyen-Stevens reaction

   $RCONHNHSO_2Ar \longrightarrow RCHO + N_2 ArSO_2^{\ominus}$

   Search tools available: CAS ONLINE-CA File, CRDS, SCISEARCH, Merck Index.

DATABASES AVAILABLE FOR REACTION SEARCHING

CAS ONLINE

The CAS ONLINE System (Registry File and CA File) can be used to search Chemical Abstracts (1967 to date) for specific compounds, generic structures or function group transformations.

Specific compound search. The family search or the complete name/molecular formula search routines are available with the Registry File to retrieve specific compounds, salts, isomers etc. The structures plus up to the ten latest references can be displayed. If only references to preparations are required, the resulting registry numbers can be automatically transferred for processing in the CA File using /P as a qualifier.

Registry File

e.g. L4 S L3 FAMILY FULL — registry numbers of compound L3 and its salts/isomers

File CA

L5 S L4/P — references to the preparation of compound L3

L6 S L4 — all references to compound L3

Generic structure searching. The substructure search routine (SSS) in the Registry File retrieves generic structures. As above, references to the preparation of the resulting compounds can be retrieved by transferring registry numbers to the CA File.

Find references to the preparation of substructure L1.

L1:

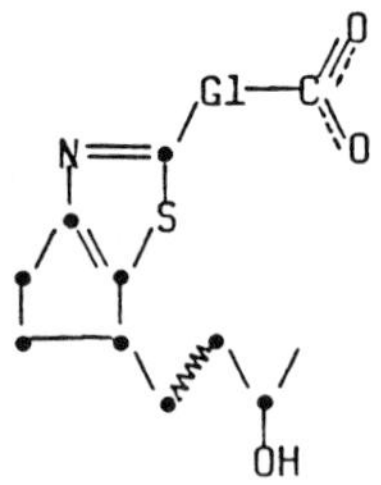

Where Gl=$(CH_2)_{1-10}$

Registry File

Set up substructure L1

L3 S L1 SSS FULL

CA File

L4 S L3/P references to the preparation of substructure L1.

Searching for the conversion of one generic structure to another:

(L2) ⟶ L1 above

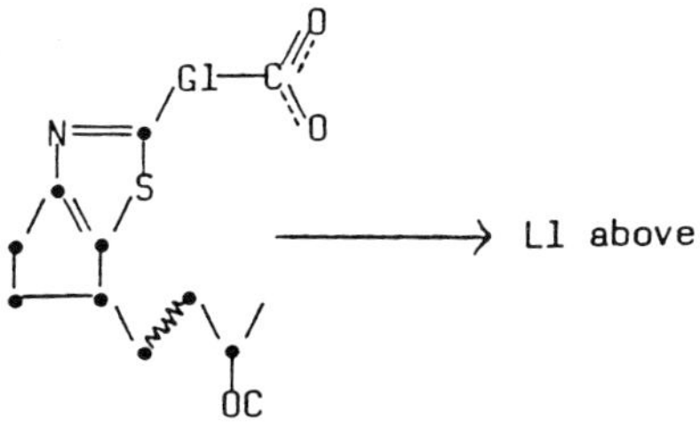

Registry File

Set up substructures L1 and L2

L3 S L1 SSS FULL
L4 S L2 SSS FULL

CA File

L5	S	L3/P	references to the preparation of substructure L1
L6	S	L4	references to substructure L2
L7	S	L5 AND L6	references that contain the preparation of substructure L1 and information relating to substructure L2. Not all these references will contain the conversion L2 ⟶ L1.

<u>General keyword searching</u>. The CA File can be searched for general keywords such as common functional groups and named reactions.

e.g. 1 S AMIDES, PREPARATION/CV AND ESTERS, REACTIONS/CV

This will lead to references covering the preparation of amides and the reactions of esters but NOT necessarily the reaction of esters to give amides.

e.g. 2 S MCFADYEN(W)STEVENS

This will produce references to the named reaction.

Advantages of CAS ONLINE for reaction searching.

* Useful for generic structure as well as single compound preparation or interconversion.

* Multiple references can be obtained since there is no restriction on the number of times a given reaction with similar yields will be included (cf. CRDS below). Even low-yield reactions are included.

* Named reactions and general concepts can be searched satisfactorily.

Disadvantages of CAS ONLINE for reaction searching.

* The system limits of 20000 compounds passing from the crude screen search to the atom-by-atom search and 5000 final hits preclude the searching of simple functional group changes in the Registry File.

* There is no capability of defining stereochemical parameters within a search strategy.

SDC FILE CRDS (CHEMICAL REACTIONS DOCUMENTATION SERVICE)

This database comprises:

Synthetic Methods of Organic Chemistry Vols. 1-30 by Theilheimer. J. Synthetic Methods vols. 1 to 10 published by Derwent.

Coverage. The coverage excludes additional routine applications of known reactions already present in the system and reactions affording 'low yields' (no definition given in manuals). This means that a search of CRDS will not be comprehensive as far as the chemist's need for information retrieval is concerned.

Keyword method of searching. The two methods of accessing this database are via keywords and multipunch. The keyword method is based on a controlled vocabulary and is the easier to use of the two options. Unfortunately, controlled vocabulary has only been applied to Theilheimer vols. 21-30 and the whole of J. Synthetic Methods. This covers the period from 1967 to date but, because routine applications of reactions already known in the system are excluded, the earlier Theilheimer volumes still have to be searched. Consequently either multipunch alone or both methods of retrieval must be used to cover the whole period of the database.

The keyword method only takes account of the reactive parts of the molecule (except important unaffected groups - see below).

Keywords are available for functional groups, rings and certain other structural features e.g. $R_1COCH_2COR_2$ has the keywords: ketone and dicarbonyl-1,3. Thematic groups, reaction terms and bond formation/breakage Theilheimer Symbols are also available for searching. Identifiers used are: /SM (starting material), /PR (product), /RT (reaction term), /TG (thematic group).

Multipunch method of searching. The multipunch method is adapted from an 80 column punch card system with each column having 10 positions. The whole structure of the starting material and product (not just the fragments taking part in the reaction) are transformed into multipunch codes.

Columns 1-27: starting material in reaction
Columns 28-54: product of reaction
Columns 55-62: bond breakage and bond formation during reaction
Columns 64-73: auxiliary substances

The different codes are 'linked' together so that only one starting material and one product occur on each 'card'. Any number of 'cards' can cover different reaction steps or isolated products under a given accession number.

Comments. In our experience, there are often many 'false drops' during an on-line search of CRDS. These must be eliminated either by looking at the index terms and title on-line (for Theilheimer vol. 21 onwards) or more effectively by checking the hard copy abstracts. It is not unusual to have to check over 100 abstracts in this way to find only approximately 20% are relevant.

Examples of 'false drops'

Multipunch

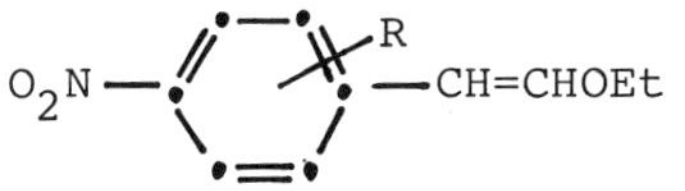

O_2N—(benzene ring, R)—CH=CHOEt would have the same set

of multipunch numbers as EtO—(benzene ring, R)—$CH{=}CHNO_2$

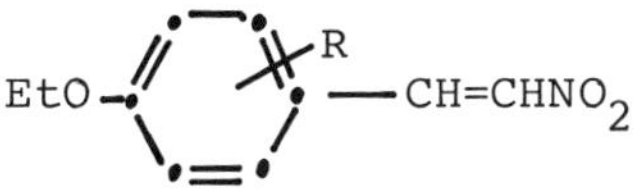

Keywords: The terms for starting material could be in the same molecule or different starting materials reacted together.

e.g. CH_3CO_2H + $CH_2{=}CH_2$ as 2 starting materials in one reaction and $CH_3CH_2{=}CH_2{-}CH_2CO_2H$ as starting material in another reaction are both covered by the index terms: C-ACID/SM AND ETHYLENE/SM.

Difficulties arise because bond breakage and formation multipunch codes have to take into account the mechanism of the reaction, not just the apparent change from one functional group to another.

e.g. $R-\overset{O}{\overset{\|}{C}}-OMe \longrightarrow R-\overset{O}{\overset{\|}{C}}-OH$

Bond broken: $\overset{O}{\overset{\|}{C}}\not-O$ <u>not</u> <u>$O\not-Me$</u>

Bond formed: $\overset{O}{\overset{\|}{C}}-O$ <u>not</u> $O-H$
↑ ↑

Problems can arise in determining the controlled vocabulary keywords for a structure.

e.g. $R_1\overset{O}{\overset{\|}{C}}NHNHSO_2R_2$

where R_1 and R_2 = carbon atom groups

Keywords = N-ACYL : $(H)C-\overset{O}{\overset{\|}{C}}-NT$ (where T = heteroatom)

HYDRAZIDE-C : $\overset{Y}{\overset{\|}{C}}-N-N$ (where Y = heteroatom)

N-SULFONYL : $C-\underset{\underset{O}{\|}}{\overset{\overset{O}{\|}}{S}}-NT$ (where T = heteroatom)

The CRDS file does have the facility, using keywords, to include non-participating groups in the reaction.

e.g. R-substituted benzene ring $-CH=CH-NO_2$ ⟶ R-substituted benzene ring $-CH_2CH_2NO_2$

Keywords: ETHYLENE-N/SM AND NITRO-C)/PR AND HC-A-CC/SY AND R/TG
↑
unaffected NO_2 group

<u>Advantage of CRDS</u>

* Excellent database for functional group interconversions.

Disadvantages of CRDS

* Too many 'false drops' that need to be eliminated by consulting the hard copy abstracts.
* Unsuitable database for specific compound or generic structure searching.
* Finding suitable keywords to use can be perplexing.
* Bond breakage and formation multipunch codes can be difficult to ascertain.
* Keywords have only been added from Theilheimer vol. 21 onwards necessitating the use of multipunch to cover the whole file.
* Structures cannot be input graphically.

THEILHEIMER ON REACCS

The tapes of Theilheimer are available for use with REACCS, via the MACCS system from Molecular Design Ltd. This is a more useful way to search Theilheimer than CRDS since REACCS/MACCS can be searched graphically. Currently this is only available as an in-house system and beyond the scope of this paper.

DIALOG SCISEARCH FILES (on-line ISI Science Citation Index 1974 to date)

These files on DIALOG are useful for looking at citations to key references for given reactions e.g. seeking references involving the McFadyen-Stevens reaction. One determines the first reference to this reaction by consulting the section on Organic Named Reactions in the Merck index 10th Edition. This is then used for the citation search.

ISI-INDEX CHEMICUS (on-line via Telesystemes Questel)

The ISI-IC database (1962-date) was recently made available on the DARC system and therefore can be searched substructurally. In theory, it should be possible to search for two compounds in a reaction sequence (A → B) and obtain references that refer to both. These would include the required references to the conversion A → B plus inevitably some 'false drops'. The fact that ISI policy is to index only compounds that are new to their system means that a novel product may be indexed without including its known starting material. Index Chemicus is therefore not comprehensive in its coverage of reaction data but does possess other useful features for satisfying the information needs of synthetic chemists e.g. the tag ED (if the paper contains full experimental details), EX (if the author reports an explosive reaction), IM (if details of instrumental methods of analysis are given in the paper). These useful features, however, do not cover the whole period of the database.

RINGDOC

One might suppose that RINGDOC via SDC could be searched for chemical reactions since the same multipunch system as in CRDS is used for all compounds. Ringdoc however, is essentially a biomedical database so that the majority of papers contain biological information. Compounds will be indexed but it is difficult to limit to chemical information only (a thematic group C for chemistry is available). It is possible to search for the existence of papers containing both types of compound in a required transformation and limit with the thematic group C. However, many 'false drops' occur making the database not very suitable to the needs of the bench chemist.

SYNLIB (Synthesis Library Software)

This is a database, marketed by SK&F, which contains reactions involving approximately 23,000 compounds. They hope to expand this to 50,000 compounds and then to issue regular six-monthly updates. The content of the database has been carefully chosen from the primary chemical literature and is searchable by entering chemical structures with the aid of a 'light pen'.

Generic or narrow searches can be carried out by selecting the key atoms of the input structure. The more key atoms defined, the narrower the search. Each entry on the database consists of reactant and product structures, yield, number of synthetic steps, reagents, reaction conditions, non-allowed coexisting functional groups and literature citation. Searching can be limited to products or reactions. Reagents can be searched for separately.

The whole database can be bought for mounting on internal computer facilities or be searched via a direct telephone line after the payment of an annual subscription fee.

LHASA (Logic and Heuristics Applied to Synthetic Analysis)

This is an interactive computer program designed to assist bench chemists in planning syntheses of complex organic molecules. The program is the most recent version of a computer-assisted synthetic analysis program which has been under development at Harvard University (and lately at Leeds University) since the early 1970s. Access to the program in this country is only through membership of LHASA UK.

LHASA is an interactive program which uses computer graphics to communicate structural information from user to computer and vice versa. The user inputs the desired molecule to be synthesised (the target molecule) and the system then notes the presence of significant structural features. The program will then generate possible synthetic precursors to the target molecule and display them graphically. The user may select one or more of these precursors for further targeting until a readily available precursor is located. At this stage one or more complete synthetic routes will have been generated.

It is important to note that the LHASA program operates in a retrosynthetic fashion and only contains sufficient literature references to exemplify the proposed conversions. It is then up to the user to consult other databases like CRDS or SYNLIB to retrieve a variety of reagents and conditions.

FUTURE REQUIREMENTS OF AN IDEAL REACTIONS DATABASE

From our experience there is currently no one information retrieval system capable of comprehensively handling the full range of reaction queries that we commonly encounter.

At present, neither publicly available online files nor traditional printed secondary sources provide an adequate index to the wealth of detailed reaction information that is presented in the primary chemical literature.

We are frequently obliged to use a range of sources in our attempts to satisfy the needs of our laboratory chemists for reaction data, yet cannot always guarantee providing a comprehensive answer from a high recall of source material. Precision of retrieved information from the online systems is often very poor.

The field of detailed practical reaction information is one where the scientist is forced to rely heavily on his past learning, experience and reading. A situation exists where current awareness services, personal data files and invisible colleges are frequently of greater value to him than retrospective literature searches.

An ideal database requires the following features:

* Available via international host(s).

* Extensive coverage of journals, patents and books.

* Graphic structure input.

* Full structure, substructure and generic search capability.

* Possibility of defining stereochemistry.

* Possibility of defining reaction site(s).

* Possibility of defining bonds being formed or broken.

* Ability to define starting materials, intermediates and products.

* Ability to handle multi-step reaction sequences.

* Searchable fields to include catalysts, reagents and solvents.

* Coverage to include low yield and failed reactions.

* Numeric data such as temperatures, pressures and yields to be included.

* Small and large scale reactions (processes) to be included.

* Registry numbers to be searchable.

* Named reactions.

* Graphic structure output of reaction sequences.

* Bibliographic citations to be searchable and displayable.

* The option of retrieving either a single example of a reaction with specified reagents and conditions or all examples of methods and conditions achieving a specified transformation.

Any database falling short of these requirements will not provide an adequate index to the huge reservoir of practical reaction details that exists in the primary chemical literature.

Table 1. Examples of printed sources of reaction information.

SOURCES	CONTENT	REACTION RETRIEVAL POINTS
J. Organic Chemistry	Full-text papers with detailed experimental sections.	Annual, Keyword index
Synthesis (Int. J. of Methods in Synth. Org. Chem.)	Major reviews (some experimental data). Full-text "communications" (some experimental and physical data). Abstracts from primary journals.	Annual, Compound index Reaction index Index of special reagents Index of apparatus and laboratory techniques
Synthetic Communications	Full-text with "sufficient exptl. detail to permit reactions and sequences to be repeated by reasonably skilled laboratory workers".	Annual, Subject index
J. Synthetic Methods	Abstracts (brief) with a single diagram e.g. from primary paper. Uses reaction symbols to indicate reaction type and bond changes.	Annual, Subject index
Current Chemical Reactions	Abstracts (brief) and flow charts.	Annual, Subject index

TABLE I (Continued)

SOURCES	CONTENT	REACTION RETRIEVAL POINTS
Current Abstracts of Chemistry	Covers novel compounds. Full flow charts from primary paper. Indicates exptl./analytical techniques used.	Annual, M.F. index Subject index Biological activities index Labelled compound index Rotaform index
Beilstein	Comprehensive literature coverage. Brief exptl. details with physical data and more recently, individual structures, shown.	Cummulative indexes Volume indexes - M.F. - Subject
Chemical Abstracts	Earlier abstracts were informative with some experimental details. Recent ones are only indicative.	Semi-annual and Collective-Period indexes - M.F. - General subject - Chemical substance
Reagents for Organic Synthesis (Fieser & Fieser)	Indicative text with flow charts and literature references.	Index to reagents grouped under a reaction type subject index.
Compendium of Organic Synethetic Methods (Harrison & Harrison)	Flow charts and literature references.	Examples arranged under functional groups.

QUERY	INFORMATION SOURCE		
	CAS ONLINE (Registry File: 1965 - date CA File: 1967 - date)	CRDS (1942 - date)	OTHER
1. Preparation of $ClCH_2C(Ph)(OH)CH_2Cl$	Registry File L1 S BENZENEMETHANOL, .ALPHA., .ALPHA.?/CN L2 S C9H10CL2O L3 S L1 AND L2 CA File L4 S L3/P Result=2 relevant hits, no false drops	Not suitable for specific compound searching	DATASTAR: Use dictionary file CNAM to retrieve reg. numbers then bibliographic file CHZZ to retrieve references. DIALOG: Use dictionary files (6 in all) to retrieve reg. numbers then transfer to bibliographic files (5 in all) using the .maprn/p facility (preparation only) BEILSTEIN This compound can be search via the molecular formula indexes.
2. $ClCH_2C(=O)CH_2Cl \longrightarrow ClCH_2C(Ph)(OH)CH_2Cl$	Registry File L1 S BENZENEMETHANOL, .ALPHA., .ALPHA.?/CN L2 S C9H10CL2O L3 S L1 AND L2 L4 S 2-PROPANONE, 1,3-DICHLORO?/CN L5 S C3H4CL2O L6 S L4 AND L5 CA File L7 S L3/P AND L6 Result=1 relevant hit (retrieved by query 1 also)	Not suitable for specific compound tranformation	DATASTAR As above. DIALOG As above but transfer registry numbers of product using /P and starting material without the /P using the .maprn facility. BEILSTEIN As above.

Table 2. Comparison of the use of different sources for handling the common types of reaction query.

QUERY	INFORMATION SOURCE		
	CAS ONLINE (Registry File: 1965 - date; CA File: 1967 - date)	CRDS (1942 - date)	OTHER
3. RCHO $\xrightarrow{NaBH_3CN}$?	Registry File L1 S CH3BN.NA/MF L2 S BORATE'('1-')', '('CYANO-C')' TRIHYDRO?/CN L3 S L1 AND L2 CA File L4 S L3 AND ALDEHYDE? Result=21 hits (14 relevant; none picked up by CRDS)	Query restricted to the specific reduction of aldehydes with $NaBH_3CN$ to give alcohols:- RNM LINK TO + NEST [ ] SS1 230+237+[450 OR 451 OR 452] + 566+610+707+710+704 SS2 ALDEHYDE/SM AND ALCOHOL/PR AND SODIUM-TRIHYDRIDOCYANOBORATE)/RT SS3 1 OR 2 Result=3 hits (1 relevant: not picked up by CAS ONLINE)	<u>Reagents for Organic Synthesis</u> <u>Fieser F. and Fieser L.F. (Editors)</u> Under sodium cyanoborohydride:- 10 references (all from 1970 onwards)
4. 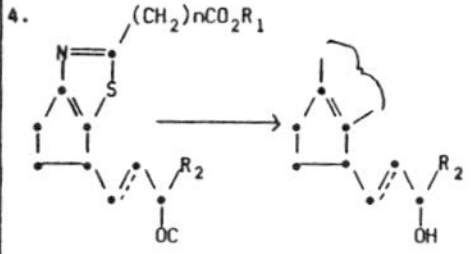 $(CH_2)nCO_2R_1$ ... R_2 OC ⟶ ... R_2 OH where R_1=H or C R_2=anything (including H) n =1-10 Further substitution is allowed.	Registry File Set up substructure L1:- 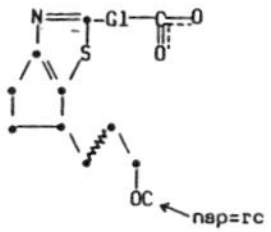 N=•-G1—C=O ... S ... OC ← nsp=rc REP G1=(1-10)· CH2 Set up substructure L2:- 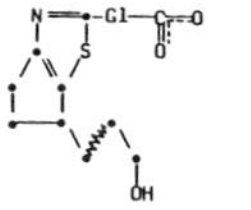 N=•-G1—C=O ... S ... OH REP G1=(1-10) CH2 L5 S L1 SSS FULL L6 S L2 SSS FULL CA File L7 S L6/P AND L5 Result: 2 relevant hits	Not suitable for this type of query.	The DARC system on Questel can be used to get similar results to CAS ONLINE.

Table 2 (Continued).

<table>
<tr><th rowspan="2">QUERY</th><th colspan="3">INFORMATION SOURCE</th></tr>
<tr><th>CAS ONLINE (Registry File: 1965 - date
CA File: 1967 - date)</th><th>CRDS (1942 - date)</th><th>OTHER</th></tr>
<tr><td>5. $CO_2R_1 \longrightarrow CON(R_2)(R_3)$

where R_1=alkyl
R_2 & R_3=H or alkyl</td><td>Registry File
Impossible to search due to system limits.
CA File
L1 S AMIDES, PREPARATION/CV
AND
ESTERS, REACTIONS/CV } 1972-date
L1 S AMIDE(L)PREPARATION
AND
ESTERS(L)REACTIONS } to cover 1967-1971
Result=34 hits (23 relevant)
N.B. Typical example of a " false drop" is a paper where amides are prepared from nitriles and acids prepared from esters.</td><td>RNM LINK TO +
SS1 234+237+505+507+(53& OR 53-)
SS2 1+564+60-
SS3 C-ESTER/SM AND C-AMIDE/PR AND NC-X-O/SY
SS4 2 OR 3
Result=212 hits (53 relevant)
N.B. Typical examples of "false drops" are where the ester is in another part of the starting material or where a hydroxamic acid is formed (not an amide)</td><td><u>Compendium of Organic Synthetic Methods</u>
<u>Vols I-V</u>
<u>Various Editors including Harrison & Harrison</u>
A total of 23 references to the preparation of amides from esters were given (12 pre-date CAS)</td></tr>
<tr><td>6.
$(R_1)C_6H_4$-$CH{=}CHNO_2 \longrightarrow (R_1)C_6H_4$-$CH_2CH_2NO_2$</td><td>Registry File
Set up substructures L1 and L2:-
L1: C_6H_5-CH=CH-$N(=O)O$
L2: C_6H_5-CH_2-CH_2-$N(=O)O$
L5 S L1 SSS FULL
L6 S L2 SSS FULL
CA File
L7 S L6/P AND L5
Result=25 relevant hits, no false drops.</td><td>RNM LINK TO +
NEST [ ]
SS1 151+157+162+17-+174+[03& OR 03- OR 030 OR 031]
SS2 386+444+393+38-+[30& OR 30- OR 300 OR 301]
SS3 1+2+551+599
SS4 ETHYLENE-N/SM AND NITRO-C)/PR AND HC-A-CC/SY AND R/TG
SS5 3 OR 4
Result=9 hits (1 exact - also picked picked up by CAS ONLINE + 8 related hits)</td><td></td></tr>
</table>

Table 2 (Continued).

QUERY	INFORMATION SOURCE		
	CAS ONLINE (Registry File: 1965 - date CA File: 1967 - date)	CRDS (1942 - date)	OTHER
7. **McFadyen-Stevens Reaction:-** $R-\overset{O}{\overset{\|}{C}}NHNHSO_2Ar \longrightarrow RCHO+N_2+ArSO_2^{\ominus}$	CA File L1 S MCFADYEN(W)STEVENS **Result=13 relevant hits**	RNM LINK TO + NEST [ ] SS1 235+237+26&+264+201+207+216+210 SS2 1+500+507 SS3 N-ACYL/SM AND HYDRAZIDE-C/SM AND ALDEHYDE/PR SS4 MCFADYEN-STEVENS)/RT OR MCFADYEN-STEVENS/TT SS5 2 OR 3 OR 4 **Result=7 hits (5 relevant)**	Merck Index, 10th Edition (1983) Under the Organic Named Reactions Section, page ONR-56, 5 references are given (2 pre-date CAS ONLINE; 1 pre-dates CRDS) DIALOG Files 186, 94, 87 and 34 (Science Citation Index 1974 - date) The key original reference for the McFadyen-Stevens reaction (ie. McFadyen J.S. & Stevens T.S., J. Chem. Soc., 1936 page 584), as given in the Merck Index was input. **Result=23 hits (9 relevant)**

Table 2 (Continued).

6 Experience with reaction indexing and searching in the IDC system

C. Fricke, R. Fugmann,
G. Kusemann, T. Nickelsen, G. Ploss and J. H. Winter
Hoechst AG

The effectiveness and survival power of any indexing language largely depends on a well planned partitioning of the tasks between vocabulary and the syntactic (synthetic) devices in this language, its grammar. The most prominent of this 'analytico-synthetic' principle is the topological approach to structure representation. Here, the periodic table of the chemical elements is taken as the vocabulary, and the various bonds between the atoms in a molecule constitute the grammar of this 'language'. Some fragment codes for molecular structure are, although more latently, based on this principle, too. This holds true especially for the GREMAS code (1-5). It is used in IDC (Internationale Dokumentationsgesellschaft fur Chemie mbH) for both structure and reaction documentation. For the purpose of reaction documentation, three types of fragments are used and three syntactic devices have been developed, each tailored for the representation of chemical reactions.

In the GREMAS system, fragments are defined and delimited in a way that conforms with the thinking of the chemist working in synthesis, who tends to think more in terms of functional groups and chains and rings than in terms of individual atoms considered in isolation. These fragments are labelled with hierarchical alphanumerical codes each consisting of three characters. In Figure 1 some examples are given for a kind of fragments which are determined by one and only one carbon atom at a time, the environment of which is included in the definition. These fragments and their codes also express the number of bonds which the carbon atom has in common with heteroatoms. This number is, by and large, identical with the oxidation state of the carbon atom. The fact is expressed, whether this carbon atom is a member of a chain or a ring and what type of ring. Other fragments which are in use for the reaction documentation in IDC are those of rings and of pairs of interconnected heteroatoms.

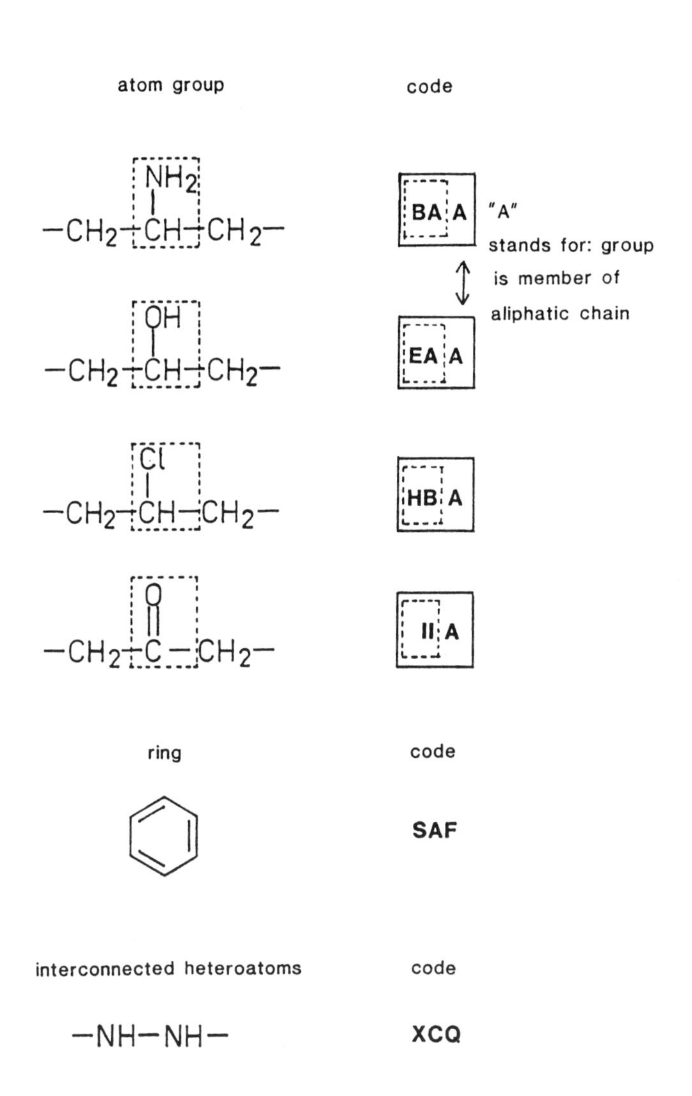

Figure 1. Examples of GREMAS codes.

When a molecule reacts, some change occurs in a fragment of the kind previously mentioned, and this fragment is then understood to be a reaction site. This change can logically be expressed as the transition from one fragment to another, and can easily be described by combining both fragment codes to yield a reaction term. An appendix to such a reaction term indicates whether or not at the reaction site a carbon-carbon bond has been cleaved or newly formed, or whether a ring structure has undergone cleavage or formation. Some examples are depicted in Figures 2 and 3.

The freedom to combine any initial and final state of a reaction site in a reaction term is one of the syntactic devices in the reaction documentation system of the IDC. As far as the recognition of reaction sites is concerned, the rule has been laid down that at least one bond must be broken or newly formed at the carbon atom in question or at one of the heteroatoms connected with it.

In most cases, several reaction sites are involved in a reaction. In order to express their association in one and the same reaction step they are combined to form a common storage unit. After that, the various reaction sites that may have been recorded in a document for separate reaction steps cannot be confused with each other in the search. In the same storage unit, catalysts and reaction conditions pertaining to the same reaction step, are also represented. This second syntactic device is necessary to avoid excessive noise of irrelevant responses that would otherwise occur in a large reaction file. Of course, in addition to reactions, the complete structures of the reacting compounds are also available for retrieval. Thus, the responses to a reaction search can be restricted to those in which certain substances or substance classes co-occur.

When searching for a certain chemical reaction one can hardly be certain that it has been described as a single-step reaction, and entered into the file in the manner previously described. Even when it is agreed among the indexers that transition states such as ions and radicals should be excluded as intermediates from the file, there is still a great variety of different pathways from a given educt to a given product. Search capabilities are therefore needed for finding reactions without having to commit oneself to the possible intermediates between an enquirer's educt and product.

With regard to this aim let us consider the storage of a multistep reaction (Figure 4). Again, for each reaction step a storage unit of its own is established. Problems emerge, however, when there are several reaction sites and/or when there are several independent reaction sequences in the document. Under these circumstances it is not clear which of the various initial and final states are related with each other in that they belong to one and the same reaction site. If this information is not preserved, one

$$-CH_2-CH(OH)-CH_2- \longrightarrow -CH_2-CH(Cl)-CH_2-$$

EAA HBA

EAA HBA - -

Figure 2. Encoding of a reaction with one reaction site.

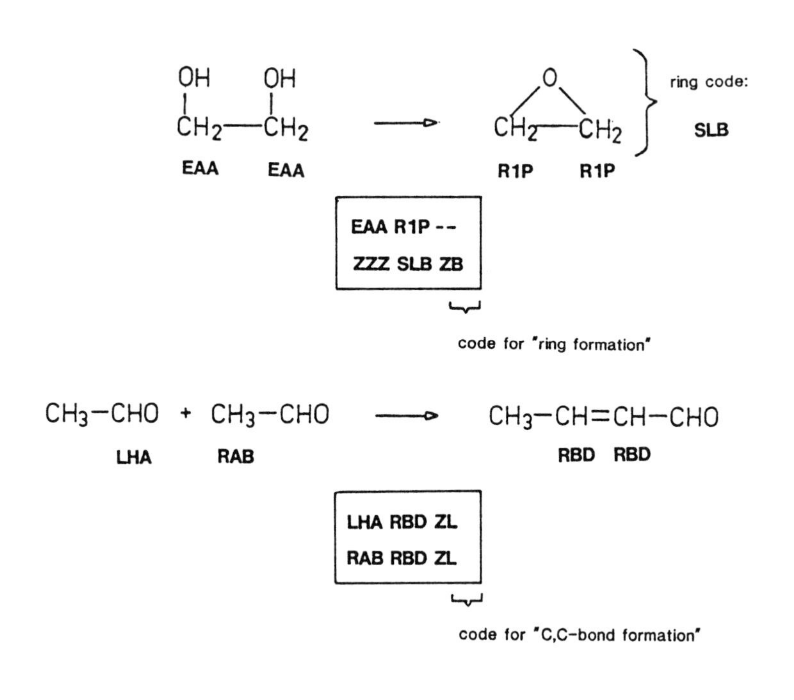

Figure 3. Encoding of reactions with more than one reaction site.

must content oneself with requiring the mere co-occurrence of, for example, a nitrile group as an initial state and of an acyl chloride as the final state of some reaction site. These very general search parameters would be satisfied, for example, by a document in which the transformation of a nitrile group into a primary amino group is reported, and, in another reaction in the same text, the transformation of a methyl group into the acyl chloride group (Figure 5). For sufficiently accurate searches it is therefore necessary to express in the query that it has to be one and the same carbon atom that fulfills the requirements of a particular initial and final reaction state. Thus requirements must be effective as a search parameter. Otherwise, many irrelevant responses would be retrieved from a large file.

In the GREMAS indexing system an arbitrarily determined identification number is assigned to each reacting carbon atom. This identification number is retained in the description of all the various reaction states of this carbon atom recorded in a document. In other words, when a carbon atom passes through several steps described in a document, the same identification number is always assigned to it (Figures 4 and 6). The search program compares the identification numbers of the reacting carbon atoms found and accepts as a response to a query only those which were recognised as belonging to one and the same reaction site. This is obvious from the recurrence of the identification number.

Only through a syntactic device like this can multistep reactions be searched with an adequate degree of precision. Figure 7 shows an example of a search that can accurately be carried out only through the employment of all three syntactic devices described earlier. The precision with which reactions can be searched in a GREMAS file comprising several hundred thousands of reactions has hitherto proved quite adequate. At least for the near future and merely for enhancing search precision topological search techniques are not yet necessary. When searching, for example, for the transformation of an olefinic double bond into a glycol structure, one will exclusively require the transformation of the olefinic carbon atom into the hydroxylated carbon atom. At the same time one will negate all other kinds of transformation of a carbon atom in this reaction step. This will exclude, for example, the hydration of the olefinic double bond (Figure 8). Here another olefinic carbon atom is changed into a saturated carbon atom, a change which was expressively excluded in the query and which correctly leads to the rejection of this reaction by the search program.

The search program also provides the possibility of negating only certain transformations in a reaction step, for example ring closure or a simultaneous rearrangement.

Only in passing we mention here that in the case of the hydroxylation of an olefinic double bond one is well advised

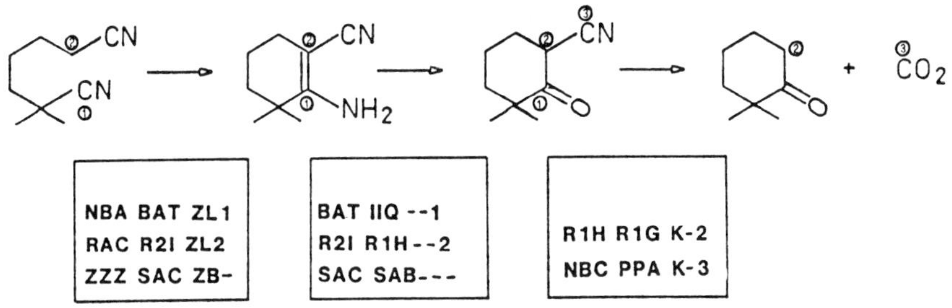
CN
CN
CN
NH2
CN
CO2
NBA BAT ZL1
RAC R2I ZL2
ZZZ SAC ZB-
BAT IIQ --1
R2I R1H--2
SAC SAB---
R1H R1G K-2
NBC PPA K-3

Figure 4. Multistep reaction with identification numbers of carbon atoms.

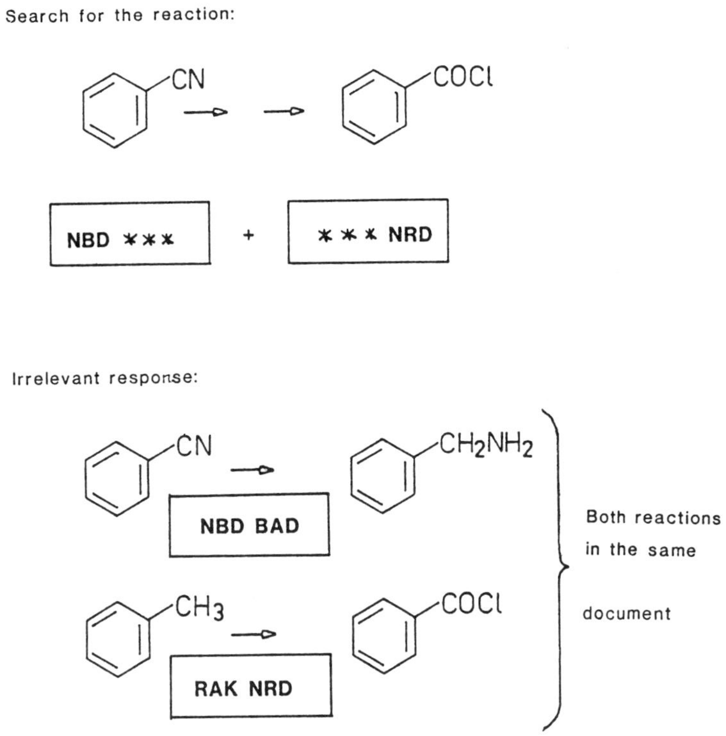
Search for the reaction:
CN
COCl
NBD ***
+
*** NRD
Irrelevant response:
CN
CH2NH2
NBD BAD
CH3
COCl
RAK NRD
Both reactions
in the same
document

Figure 5. Irrelevant response to a generalised search due to lack of synthetic devices.

SEARCH: NBD *** + *** NRD

Requirement: In both cases the same identification number

Correct responses:

Document A:

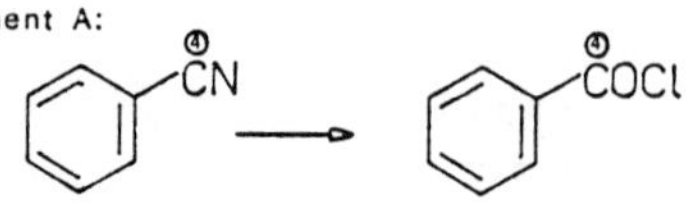

Document B:

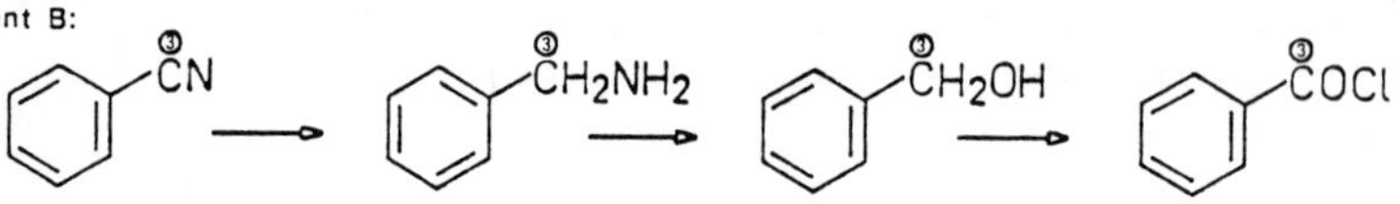

Document C:

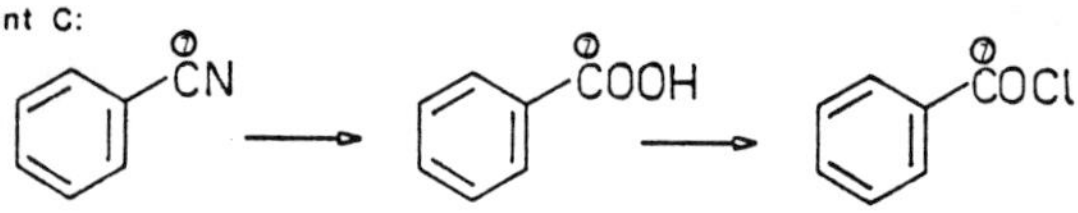

Figure 6. Correct responses for generalised search with identification numbers as suitable syntactic device.

Searched is the following sequence:

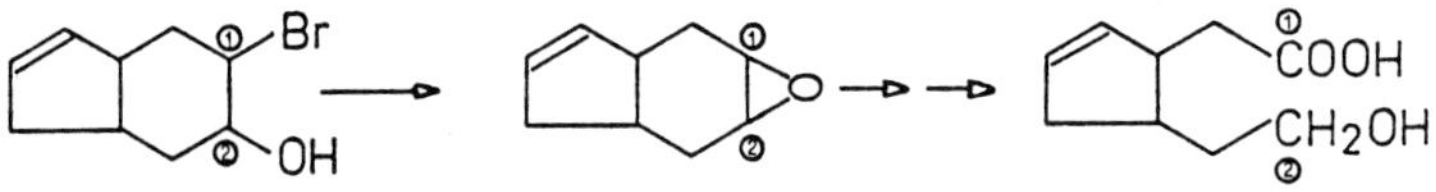

Search codes:

1.)	HCQ R1Q	+ *** NNA	
2.)	EAQ R1Q		+ *** EAA
	ZZZ SLB ZB		

Requirements: Codes in line 1.) must have the same identification number.
The same must be the case for the codes in line 2.) .

Figure 7. Reaction search with three syntactic search parameters.

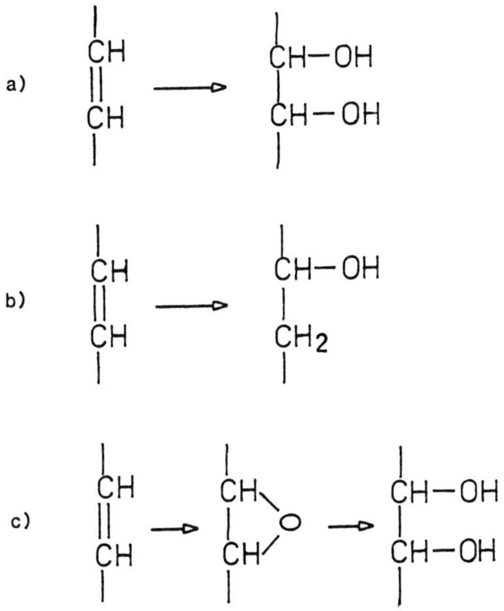

a) $-CH{=}CH- \longrightarrow -CH(OH){-}CH(OH)-$

b) $-CH{=}CH- \longrightarrow -CH(OH){-}CH_2-$

c) $-CH{=}CH- \longrightarrow -CH{-}CH-$ (epoxide, bridged by O) $\longrightarrow -CH(OH){-}CH(OH)-$

Figure 8. Three similar reactions with different search requirements.

to permit multistep reactions. Otherwise, one would, certainly undesirably, exclude reactions in which, in a first step, an epoxide ring was formed, and, in a second step, this ring was cleaved to yield the desired glycol structure.

The purely manual indexing of reactions using fragment codes of the kind previously described entails a lot of work. It is therefore desirable to mechanise this work as far as possible. One way of doing this, which was recently investigated in HOECHST, is to enter the educts and products by using conventional topological input routines and to identify the reaction sites manually. Afterwards, machine programs are employed for the conversion of this input format into the fragment code representation. Another advantage of this approach is that the educt and product structures are then made available for precise topological search (Figure 9).

REFERENCES

1. R. Fugmann, W. Braun and W. Vaupel, 'GREMAS - ein weg zur Klassifikation und Dokumentation in der organischen Chemie'. Nachrichten fur Dokumentation, 14, 179-90, 1963.

2. R. Fugmann, W. Bitterlich, 'Reaktionendokumentation mit dem GREMAS - System (Reaction Documentation using the GREMAS System)', Chemikerzeitung, 96, 323-30, 1972.

3. R. Fugmann, 'The IDC System', in Chemical Information Systems, edited by Janet E. Ash and Ernest Hyde, Ellis Horwood Limited, pp. 195-226, especially p.213.

4. R. Fugmann, G. Kusemann and J.H. Winter, 'The supply of information on chemical reactions in the IDC System', Information Processing & Management, 15, 303-23, 1979.

5. S. Rossler, A. Kolb, 'The GREMAS System, an integral part of the IDC system for chemical documentation', Journal of Chemical Documentation, 10, 128-34, 1970.

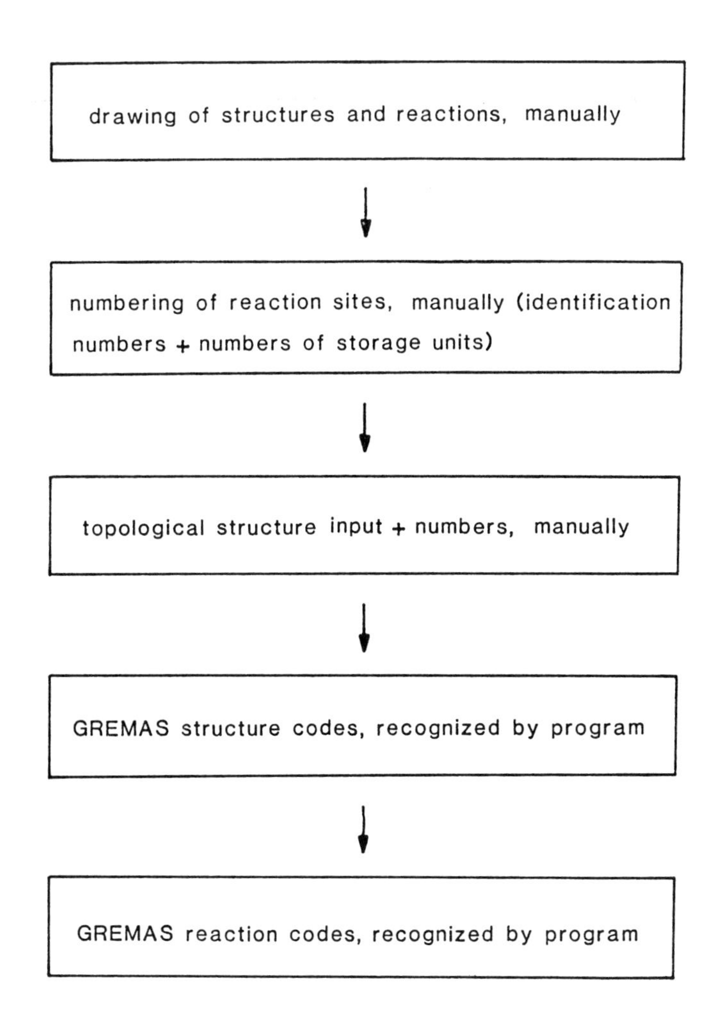

Figure 9. Partially mechanised structure and reaction input.

7 Design, implementation and evaluation of the CONTRAST reaction retrieval system

D. Bawden and S. Wood
Pfizer Central Research

SUMMARY

The CONTRAST reaction analysis system, developed in-house at Pfizer (UK) as a component of the SOCRATES chemical information system, is described. The rationale for the system, and its design constraints, are noted, and its main components outlined. In particular the automatic reaction analysis techniques, screening procedures, and query language, are described. Finally, the integration of CONTRAST into an overall system for computer-aided organic synthesis is outlined.

INTRODUCTION

CONTRAST (COnnection Table Reaction Analysis and Search Techniques) is a computerised system for analysis, storage, and retrieval of chemical reactions information, developed at Pfizer Central Research (UK). It is designed as a component of the SOCRATES chemical and biological data-bank system, also developed at Pfizer Central Research (UK).

The purpose of this paper is to describe the CONTRAST system, emphasising those aspects which particularly distinguish it from other reaction retrieval systems. Particular emphasis will be given to the rationale for the overall system design, and its integration with the SOCRATES structure searching system. The components of CONTRAST which merit detailed discussion in this paper are the automatic reaction analysis procedure, the screening process, and the searching language and facilities. The role of CONTRAST as part of a set of computerised aids to organic synthesis will be described, as a conclusion.

SYSTEM DESIGN

CONTRAST is a reaction analysis and search system designed as an integrated adjunct to a structure searching system SOCRATES. A principal requirement was that as much as possible of the existing SOCRATES software should be used in CONTRAST, to minimise the effort necessary in building, and subsequently maintaining, the system. A consequent requirement is that CONTRAST, like SOCRATES, is written in standard FORTRAN 77 to run on VAX computers, using VT640 or VT240 graphics terminals. The query language for reaction searching should have as much in common with SOCRATES structure searching as possible.

The SOCRATES system (1) will not be described in detail here. Its chemical data-bank modules store and search redundant connection table representations of chemical structure, with an interactive graphics query input, and substructure searching by a two stage process of fragment screening and atom-by-atom searching. Its general principles are those embodied in a number of recent generation chemical information systems, such as CAS ONLINE, DARC, and COUSIN(2).

Unlike some other reaction retrieval systems, CONTRAST is not associated with any particular data-base. On the contrary, the system is able to accept any data-base having full structure representations of reactant and product molecules, regardless of the presence or absence of any reaction site information, and any additional indexing terms, data etc. Thus CONTRAST can work with reaction files constructed in-house (from internal reaction information or reactions selected from the literature) or with commercially available files, or with those constructed in collaborative projects between organisations. Large reaction data-bases of general interest and smaller 'personal files' can be handled equally well.

Allowing for this degree of flexibility in data-base creation places strong restrictions on the sort of reaction information with which the system will deal. Four important constraints should be noted:

a) formal structural changes only are considered, with no attempt made to describe mechanism.

b) the reaction analysis procedure is entirely algorithmic, and based only on reactant and product structure, with no intellectual intervention whatever (except in any case in which the algorithms fail to give a valid analysis).

c) reaction conditions, reagents, and any other numeric or textual information which may be present is stored in free-text data fields associated with the reaction, so that it may be displayed in answer to a structural search, or (at most) searched as free-text. This gives the flexibility to include information of whatever sort may be present in any data-base with which CONTRAST is used, at the expense of the sort of sophisticated access to such data, allowed by some reaction retrieval systems, e.g., SYNLIB or ORAC (3).

d) CONTRAST works with reactions comprising single-step one reactant to one product transformations. Thus multi-reactant, multi-product, and multi-step reactions must be broken down, as exemplified in Figure 1. Intellectual decisions may have to be made to select sensible reactions.

A + B ⟶ C + D

represented as some or all of

A ⟶ C
A ⟶ D
B ⟶ C
B ⟶ D

A ⟶ B ⟶ C ⟶ D

represented as some or all of

A ⟶ B
A ⟶ C
A ⟶ D
B ⟶ C
B ⟶ D
C ⟶ D

Figure 1. Representation of reactions in CONTRAST.

These constraints, together with the requirements for maximum compatibility of CONTRAST with SOCRATES in software and in the user interface, dictate that the system has four main components:

i) algorithmic reaction analysis, including procedures for automatic reaction site detection, and for the detection of ring changes and molecular formula changes.

ii) a two-stage reaction search procedure, by screening and atom-by-atom searching (exactly analogous to that in SOCRATES structure searching), with the addition of searching on other reaction parameters.

iii) a graphical query input, with as much in common as possible with SOCRATES structure search graphical system, with additional input of reaction parameters.

iv) output of search results.

These components will now be considered in more detail.

MAJOR SYSTEM COMPONENTS

Algorithmic reaction analysis

The most crucial part of the reaction analysis component is a fully automated procedure for detection of reaction sites by comparison of the connection table representations of reactant and product structures. Subsidiary routines detect changes in rings and in molecular formula, for reactant and products. This latter information is needed for carrying out very general searches (e.g. reactions forming five membered rings), and also for refining searches specified graphically.

The reaction site detection procedure is a slightly modified version of the approximate structure-matching algorithm described by Lynch and Willett (4,5), which determines equivalences between similar atoms in reactant and product structures by application of the iterative Morgan algorithm. Our modifications to Lynch and Willett's technique are largely restricted to 'recovery procedures', which enable certain 'difficult' types of reaction to be satisfactorily processed (see below).

This process generates small reaction sites, including only those atoms which are added to, or deleted from, the reactant structure, or which change their bonding patterns; thus giving a reaction site corresponding to the minimum formal structural change. It is important to note that atoms are not included in the reaction site simply because they play some significant part in the reaction, e.g. promoting or interfering functionality, (as is the case with SYNLIB (3)), nor even because they are essential for the conventional definition of the reaction (as is the case with ORAC (3)). This formal, algorithmic approach has the advantage of total objectivity, unambiguity, and reproductivity.

Subsidiary procedures generate ring change records (by a comparison of all ring systems, defined by ring size and atomic composition, in reactant and product) and overall molecular formula changes for both reactant and product (expressed as positive or negative as appropriate). As with the site detection, these analyses related to purely formal structural changes.

This reaction analysis procedure was tested on a file of 8000 reactions available to us in machine-readable form. 97% were judged to have been analysed successfully, in that they passed fairly stringent automatic checks on the validity of the results, and all those inspected appeared sensible analyses. The recovery procedures added about 5% extra analyses, to an initial 92% correctly analysed, which compares with 93% found by Lynch and Willett, using a different data-set (4). The 3% which still failed fell into three main categories:

a) small change in large molecule (52% of failures)
b) large change in small molecule (41% of failures)
c) complex multisite reaction (7% of failures)

Steps to deal with these could include intellectual intervention, treating the whole structure as the reaction site (appropriate for cases a) and b) above), specific exception conditions to deal with commonly occurring errors etc.

These results give us confidence that the analysis procedure is sufficiently robust and accurate for general use, particularly since the test file included a variety of both literature and in-house reactions, as well as examples put in specifically to examine the system's limitations.

Reaction search procedure

The first stage of the reaction search is a screening step, which is described in detail below. Those reactions which pass this stage are then matched exactly with the query by an atom-by-atom match (if a substructure has been entered graphically) and/or a detailed matching of reaction parameters (i.e. exact checking of changes in rings and molecular formula). The atom-by-atom matching uses the same software for back-tracking graph matching as in the SOCRATES structure searching system (1), with the sole addition of an indication for each atom that it is in, is not in, or maybe in, the reaction site.

The screening system for the reaction search uses 662 fragment screens, algorithmically defined and chosen on frequency considerations, in the same way as the 1315 fragment screens used in the SOCRATES substructure searching system (1). The same inverted bit-map software is used for searching, and, from a graphical substructure input, both reaction and structure screens are automatically invoked.

The fragments used as reaction site screens are chosen from the atom- and bond-centred fragments devised for substructure searching by Adamson and Lynch and their coworkers at Sheffield (6). The smaller types of fragment are used, since they are generated from that part of the structure comprising the (generally small) reaction site. The fragments used, with the numbers of different types, are listed, and exemplified, in Figure 2.

191	augmented pair e.g.	(1) C-N (1)
70	simple pair	C-N
43	atom pair	C N
67	bonded atom	-N-
37	connected atom	N(2)
16	simple atom	N

Figure 2. Reaction site screens.

Fragments describing ring changes are also used as screens. These are listed and exemplified in Figure 3. Finally 20 screens describe increases in atom counts, and another 20 corresponding decreases.

In this way, a reaction screening system can be rapidly constructed, sharing design concepts and software implementation with a structure search screen set.

69 ring size and composition
e.g. 6 N C C C C C

50 ring size and atom type
e.g. 6 N C

13 atom types in ring
e.g. N C

43 single atom type and ring size
e.g. 6N, 6C

9 single atom type
e.g. N, C

14 ring size
e.g. 6

Figure 3. Ring change screens.

Query input

The input of reaction queries has two stages, either or both of which may be used to describe reactant and/or product.

The first stage is a graphical substructure entry process. This utilises the standard SOCRATES substructure search menu, with the sole addition of menu options for specifying that atoms are or are not in the reaction site. The requirement to use just two additional options, out of a total of about forty, makes the reaction searching system particularly easy to learn, and helps in gaining acceptability with users, as well as achieving the objective of sharing software to maximum extent. The full facilities of SOCRATES substructure search (1), for inputting structural queries of varying degrees of specificity are of course available. Reaction sites may be specified for reactant and/or product, or for neither.

The second stage is the input of additional reaction parameters (ring and molecular formula changes) by a menu system - a command-driven procedure could equally conveniently be adopted for more experienced users. Specifying reaction parameters alone, without graphical substructure input, gives access to reaction information at a very general level, which may be particularly useful for overviewing an area of chemistry. Reaction parameters would be specified for only one of reactant or product, because of the symmetric nature of the analysis, i.e. a ring formed in the product is equivalent to the same ring broken in the reactant etc.

Reactant conditions, reagents, yields and similar associated data are at present simply stored in an associated data-file, to be displayed as output. It is not planned to provide more than a relatively simple free-text search capability for such associated data, because of the requirements for flexibility in accepting data-bases with very varied indexing policies.

Output

The standard SOCRATES graphical structure display routines are used for output of reaction sequences, with the addition of 'highlighting' for the reaction site if required. Any associated data is output as text on the same screen. Hard-copy is readily provided via a laser printer.

OVERVIEW OF SEARCHING

The system components described above allow a highly flexible approach to reaction searching, in terms of structural changes. Among the more important types of searches which can be undertaken are:

- reactant structure
- product structure
- reactant reaction site
- product reaction site
- generic structural change

Any or all of these may be combined, with reactant and product structures (with or without site specification) searched separately or in combination, or together with a generic specification of structural change. Substructures, in reactant and/or product, may be specified as reacting or unchanged, or may be left unspecified.

It is worth re-emphasising that searching is oriented purely towards formal structural changes, and hence mechanistic considerations, and other aspects not expressible in structural terms are not catered for. This may be overcome, to a limited extent, by provision of text searching on associated data or descriptors, which may be present in data bases used with CONTRAST.

FUTURE PLANS

Data-bases

The most obvious requirement for effective operational use of CONTRAST is to provide access to a variety of reaction data-bases of high quality. This can best be done by a combination of the means mentioned above - acquisition of commercial files, in-house compilation etc.

Similarity searching/browsing.

CONTRAST, like SOCRATES, provides a precise retrieval tool for chemical information. Experience with SOCRATES has shown the value of procedures for browsing, ranking output etc., based on calculation of quantitative measures of similarity between chemical structures (7). The same procedures, working with measures of similarity between reacting centres, should provide a valuable reaction browsing ability in CONTRAST.

Integration with other CAOS systems.

Reaction retrieval is an important part of the wider topic of computer-aided organic synthesis (CAOS). The components of an integrated approach to this area, as envisaged at Pfizer Central Research (UK) are listed below:

- structure searching (e.g. Fine Chemicals Directory)
- reaction retrieval (CONTRAST)
- reaction browsing (e.g. SYNLIB)
- synthesis planning (e.g. LHASA)
- reaction prediction (e.g. CAMEO)
- molecular orbital/molecular mechanics calculations.

The value of providing these tools in as integrated a fashion as possible, sharing as much of a common interface as is feasible, need not be stressed.

It is worth mentioning particularly the complementary nature of CONTRAST and the SYNLIB reaction information system (3). CONTRAST, as has been made clear, is a precise retrieval tool, based on algorithmic analysis of formal structural changes. SYNLIB is intellectually indexed, with a considerable degree of chemical intelligence built in, and with searching facilities designed to facilitate a browsing style of exploration of particular areas of chemistry. It is therefore a very valuable alternative approach to CONTRAST: the same may be true of other reaction systems now becoming available, such as ORAC. In general it may be desirable to have more than one searching program working on a given data-base, thus maximising the value of the information resource.

CONCLUSIONS

The development of the CONTRAST system, taking in all about six man-months, demonstrates the feasibility of devising reaction retrieval components, sharing much of the software and user interface with substructure searching systems. Such components are best based on algorithmic analyses of reactions in terms of formal structural changes. They form a valuable part of integrated approaches to computer-aided organic synthesis.

REFERENCES

1. D. Bawden et. al., paper in preparation.

2. J. Ash, P. Chubb, S. Ward, S. Welford and P. Willett, 'Communication, Storage and Retrieval of Chemical Information', Ellis Horwood, Chichester, 1985.

3. see the relevant papers in these proceedings.

4. M. F. Lynch and P. Willett, 'The Automatic Detection of Chemical Reaction Sites', J. Chem. Inf. Comput. Sci., 1978, 18, 154-159.

5. P. Willett, 'The Evaluation of an Automatically Indexed, Machine-readable Chemical Reactions File', J. Chem. Inf. Comput. Sci., 1980, 20, 93-96.

6. G. W. Adamson et. al., 'Strategic Considerations in the Design of a Screening System for Substructure Searches of Chemical Structure Files', J. Chem. Doc., 1973, 13, 153-157.

7. P. Willett, V. Winterman, D. Bawden, paper in preparation.

8 RMS-DARC Reaction Management System: a new software produced by Télésystèmes DARC

J. P. Gay
Télésystèmes DARC

SYSTEM ARCHITECTURE

RMS is a reaction storage and retrieval system which means that, using RMS you create a reaction data base you can then search to answer the questions you may have regarding synthesis problems.

In RMS, a reaction data base is an extension of a structural data base, which means any structure base available in a chemical company can be the starting point of a reaction data base.

In this data base each structure is identified by a UID (User Identification Number), a structure sheet is linked to this UID.

Each reaction is identified by a RID (Reaction Identification Number) and a reaction sheet is linked to the RID.

INPUT

According to this architecture, RMS consists of two main graphic menus for input:

- one for reaction/reaction sheet input
- one for structure/structure sheet input

To create a reaction two different techniques are possible:

- if you know the UIDs of the starting materials, products, reagents or solvents, you can just recall the structure

- if the structures are new (or if you do not know the UIDs) you can draw the starting materials, products, reagents or solvents (each can contain 255 non-H atoms).

(The novelty of the structures is checked in RMS so that you can avoid duplication of a structure).

The structures are stored according to the CAS-R3 convention for aromaticity and tautomerism and you can decide either to store your drawing or not.

Of course you can combine both techniques to input up to four starting materials, six products, four reagents and four solvents.

The reacting center input is performed just by pointing at the atoms and assigning a number.

The input is checked so that the numbering of the products is consistent with the numbering of the starting materials (taking into account the stoichiometric coefficients).

The reaction sheet input, as well as the structure sheet input is performed very easily through a graphic menu containing 22 fields.

These fields are defined through a data definition language. As an example you can define: author, reference, year, chemical family, physical condition, yield, yield comment, temperature, temperature comment, pressure, pressure comment, PH, medium, solvent, catalyst.

At any time during the input you can display the reaction and the data you already put in. The reaction display includes the UIDs of the structure, the number of steps and the stoichiometric coefficients.

This is the description of the interactive input of RMS. A batch creation of a reaction data base from WLN and Connection Tables and Data files is also possible.

SEARCH

As for input, RMS consists of two main graphic menus for search:

- one for reaction search
- one for structure search

Both are based on the Generic-DARC capabilities.

The main types of queries you can phrase with RMS are:

- How can I make 4 substituted Isoquinolines?

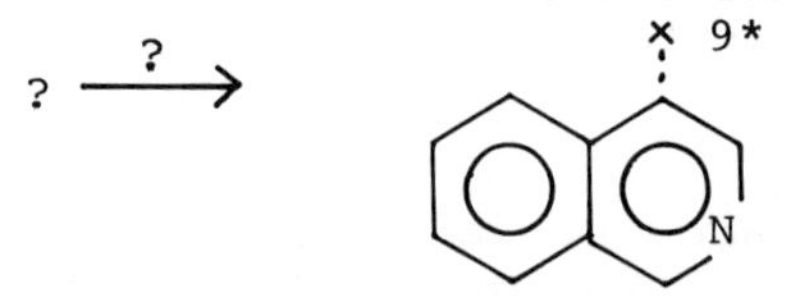

(Substructure with a substituted undefined atom and an undefined bond)

- What products can I make from 3,5 Diamino-Chlorobenzene?

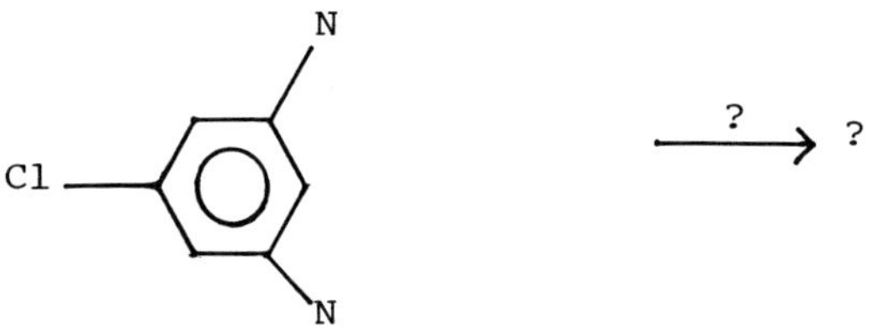

N

Cl

N

$\xrightarrow{?}$? (structure)

(structure)

- How can I cyclically protect a Cis Cyclohexane diol?

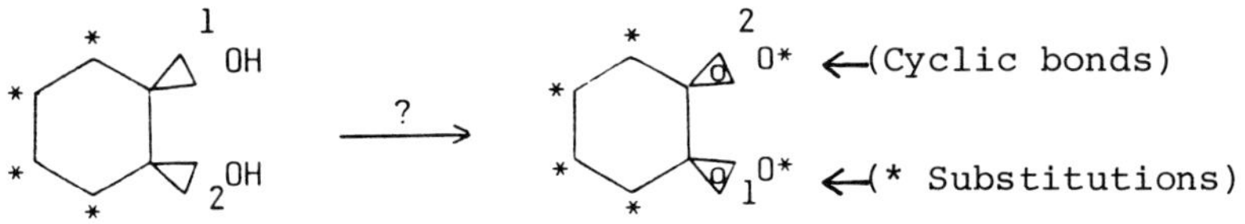

1 OH

2 OH

$\xrightarrow{?}$

2 O* ←(Cyclic bonds)

1 O* ←(* Substitutions)

(substructure) (substructure)

- How can I make a N-S bond?

$_1N_{3*}$ + $_2S_{6*}$ ⟶ $_1N_{2*}$ ... $^-{}_2S_{6*}$

(substructure) (substructure) (substructure)

(no fragment search limitation)

- How can I reduce a nitro group in the presence of a benzoic lactone?

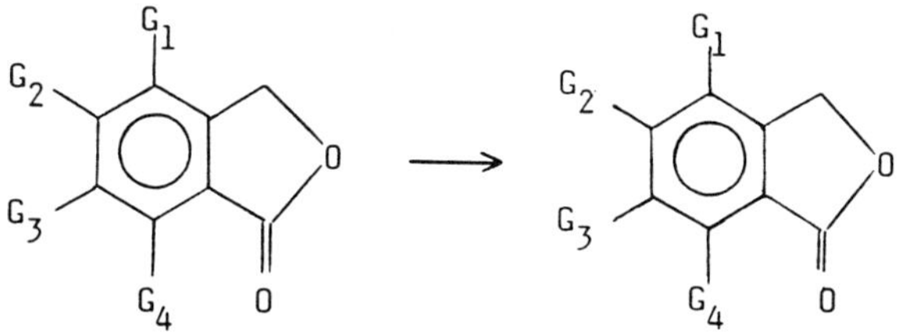

G_1 G_2 G_3 G_4 O O ⟶ G_1 G_2 G_3 G_4 O O

$G_1 = G_2 = G_3 = G_4 = H, NO_2$ $G_1 = G_2 = G_3 = G_4 = H, NH_2$

(generic structure) (generic structure)

This example illustrates the use of generic group in dealing with undefined attachments.

- How can I reduce a nitro group in the presence of an aldehyde?

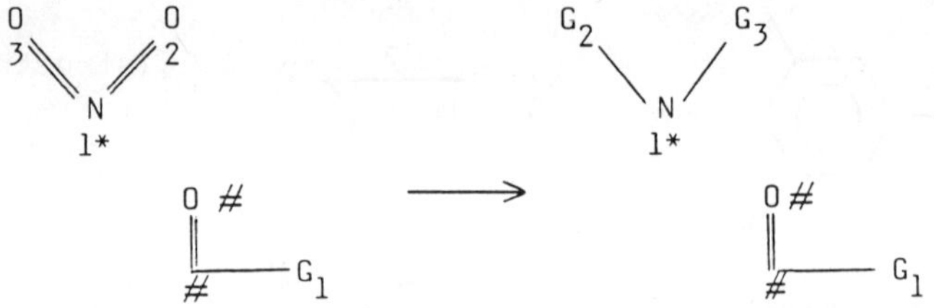

G_1 = H $G_1 = G_2 = G_3$ = H

(1, 2, 3 reacting centres, # no reacting centre)

This example illustrates the use of H atom in a generic group and the exclusion of reacting centres on certain atoms.

The use of generic may also be illustrated by several other examples such as:

- How can I create a cyclic ether from an ethylene oxido compound?

  In this case the variable size of the ring (product) will be defined through a variable Group G_1 in a 4 member ring (for instance).

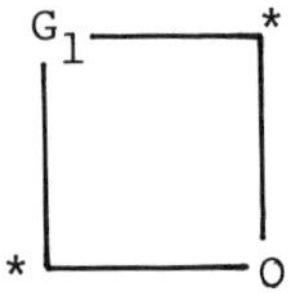

$G_1 = C_2, C_3, C_4, C_5, \ldots$

Queries such as:

What uses have been reported for the tertio-butyl lithium as reagent?

or: Do betadiketones react with methyl isocyanate as reagent and what are the products?

can also be phrased with RMS.

All the reaction searches may be refined by a data search. The data search includes:

Numerical search through the operators: =, #, >, <
Data search with right and left truncation

The boolean operators AND, NOT, OR can be used.

REACTION DISPLAY

The results of a search are displayed according to user defined report formats. (A max. of 100 formats for one data base).

A report format consists of a reaction box and a maximum of 47 data boxes, all user defined including data boxes above and under the arrow to get the usual reaction display.

A default report format is automatically selected when you ask for the display of a reaction.

Just by pointing at boxes on the display menu you can select other report formats for the current reaction, other reactions with the current format and combine both. 16 classes of security are offered to manage security problems.

9 Exploring reactions with REACCS

W. T. Wipke, J. Dill, D. Hounshell, T. Moock and D. Grier
Molecular Design Limited

ABSTRACT

This paper briefly describes the REACCS Reaction Access System and the data bases that are currently available for searching with REACCS 6.0. We show using REACCS 6.0 and these data bases that it is possible for a chemist to conveniently browse computer-readable reaction data bases. Examples of browsing Organic Syntheses and Theilheimer's Synthetic Methods are included. Stereochemical substructure searching is used to find examples of a pericyclic stereospecific reaction. We find that REACCS permits the old and the new reactions to appear adjacent in time and space, enabling the chemist to analyse them within the same context, and to see new relationships between them.

Reactions are certainly as important to chemists as are individual chemical structures, but only recently have the tools been developed to facilitate computer storage and retrieval of chemical reactions. Valls perceived the need: 'I am convinced Chemists badly need a Reaction Documentation (system) and when such a tool is put at their disposal they use it intensively ... it means that retrieving information on reactions is of major - I would say of vital importance, and at least as necessary as retrieving information on chemical compounds.'(2) That need was first filled by Molecular Design Limited with the development of REACCS™, the Reaction ACCess System, which was demonstrated publicly in 1981. (3-6) Since that time enhancement of REACCS has continued and large high quality databases have been constructed.

The purpose of this paper is to describe the latest version of REACCS, REACCS 6.0, to describe the data bases available for REACCS, and to illustrate how one can learn chemistry by browsing through these data bases.

Our goals for REACCS are summarised:

- complete representation of reactions
- representation of reaction variations
- reacting centres and complete context
- flexible, extensible, data structure
- friendly graphical communication
- powerful substructure search of reactions
- MACCS capabilities for molecules
- stereochemical searching

By complete representation we mean that all chemically significant information including complete description of all reactants, reagents, catalysts, solvents, and products be represented in a form in which information is not lost. For chemical structures, this means a connection table representation. We can represent all reactants and all products to avoid any bias about what is an important molecule in the reaction. Thus twenty years from now our representation of the reaction will still be valid and useful, even though our ideas of what is important or interesting may have changed.

Reaction variations are reactions involving the same reactants and products, but which differ in quantities, experimental conditions, yield, reference, workup, etc. The variations collect this valuable information together in one place without duplicating the reaction.

Reacting centres are the bonds broken or made in the reaction. These bonds are designated by algorithm or manually and are displayed on the screen highlighted. As we will see, search queries can designate that a bond in the query must be part of a reacting centre, or must not be part of a reacting centre, but additionally, one is not restricted to searching over reacting centres. In addition to the reacting centres, the remainder of the chemical structure is available as context.

The data structure of REACCS is hierarchical and user-defineable. This permits one to design an in-house reaction data base to fit the needs of the organisation. One application may need extensive workup data while another may be concerned with kinetic rate data.

Following the example of MACCS™, (7-9) REACCS utilises interactive computer graphics for input and output of reactions and search queries. The powerful substructure search capabilities of MACCS are available for substructure searches on both sides of the arrow. REACCS also provides for storing molecules and for MACCS-like searches over molecules. The molecules need not be involved in any reaction. And as in MACCS, REACCS permits a complete stereochemical description of each molecule in the reaction to be stored and retrieved, and also to be searched via stereochemical molecule or reaction queries.

Graphical display of reactions is also user-defined. The user specifies what data items should be displayed and the region on the screen in which to display each item. Units of measure are part of the user profile, as are choices of molecule identifiers to be displayed. In a search one can see each reaction (hit) one at a time (form view), or one can see selected datatypes displayed as a table (table view). The display of complex chemical structures can be simplified by automatic substitution of structural abbreviations such as 'Ph' for phenyl groups. The abbreviations replace the structural diagrams for that structural fragment, but REACCS retains the full structural detail within.

Reacting centres are highlighted to assist the chemist in readily seeing the parts of the structures undergoing reaction, thereby speeding the chemist's decision regarding whether the hit is 'interesting'. With large molecules (up to 256 non-hydrogen atoms) this really speeds information flow to the chemist. Reactions are displayed in the normal manner with all reactants on the left, all products on the right, and solvents, catalysts, and conditions above and below the arrow. Quantities of each chemical, or any molecular data may be displayed below each chemical. Any molecule in the reaction may be selected for full screen molecule display together with the molecular data that are available and that are specified by the user to be displayed in the molecule form.

Device independence in graphics is particularly important because graphic terminal technology changes so rapidly. The current hardware compatibility of REACCS 6.0 is shown below.

Computers	IBM, Fujitsu, Prime, VAX
Terminals	Imlac, Tek 4105 4107 4114, VT640, Lundy 568X, DQ650M, HP2623A, GT40, Envision
Hardcopy-Remote	Versatec, CalComp, HP7221, Metafile
Hardcopy-Local	Tek 4010 emulation, Tek 4662, HP7332 747XA 2623A

Through the Metafile, hardcopy is also available on the QMS Laserprinter and user-supported hardcopy devices.

Among the enhancements of REACCS 6.0 are new easy-to-use graphic user interface which you will see in the screen images, increased search efficiency, and an expanded offering of data bases: Organic Syntheses, Theilheimer Synthetic Methods, Fine Chemicals Directory (FCD), Current Reactions Literature File (CLF), and the Journal of Synthetic Methods.

Organic Syntheses currently contains approximately 4,500 reactions, each of which has extensive quantitative data, workup procedure, safety warnings, and literature references. Each reaction has been checked experimentally and the name of the checker is part of the data base, so these are among the most reproducible experimental procedures in the literature. The reactions also are selected on the basis that they are general preparative procedures. The compounds used as reactants are generally more simple than those found in other reaction data bases.

Theilheimer's Synthetic Methods of Organic Chemistry Volume 1-35 contains a total of approximately 44,000 reactions. To create this collection Dr Theilheimer, an industrial chemist, selected 1,000 to 1,500 reactions per year from 1946 to 1981. These are novel synthetic reactions selected from 250 journals. The data base includes catalysts, solvents, stereochemistry, original literature reference, Theilheimer reference, and supporting data. In creating the data base from the books, the text and nomenclature are used to check the structural diagrams for errors and to add stereochemistry where none was represented in the diagram in the book. Volumes 26-35 have been released in REACCS form. From this initial release, we have been able to examine the distribution of journals covered. The most frequently cited journals in volumes 26-35 are listed in Table 1 together with the count of citations and percentage. Clearly, these data indicate quite wide coverage of the journals since the most heavily cited journal only represents seven percent of the citations and the tenth most heavily cited journal is only at the one percent level.

Journal	Count	Percent
Journal Organic Chemistry	1,739	6.92
Tetrahedron Letters	1.609	6.32
JACS	926	3.68
Synthesis	778	3.10
Chemical Abstracts	583	2.33
Chemical Communications	532	2.12
Synthetic Methods	400	1.61
SOC Perkin I	332	1.33
Berichte	304	1.21

Table 1. Frequency of citation of journals in Theilheimer Volumes 26-35.

Similarly we have analysed the frequency of citation of authors in volumes 26-35 of Theilheimer Synthetic Methods in Table 2. The most frequently cited author constitutes less than one percent, again supporting the wide, even coverage of the literature by Dr Theilheimer.

Author	Count	Percent
T. Mukaiyama	102	0.82
E. J. Corey	71	0.57
D. H. R. Barton	50	0.40
H. C. Brown	44	0.36
F. Yoneda	41	0.33
B. M. Trost	38	0.31
E. C. Taylor	37	0.30
G. A. Olah	35	0.28
Y. Tamura	33	0.27

Table 2. Frequency of appearance of authors in Theilheimer Volumes 26-35.

Multi-step sequences in Theilheimer are represented in the REACCS data base as each individual step plus an overall step, thus A ⟶ B ⟶ C is represented by three entries, A ⟶ B, B ⟶ C, and A ⟶ C. In Volumes 26-35 of Theilheimer, as shown in Table 3, about 90 percent of the entries are single steps, and 98.5 percent are two steps or less. Consequently, the above treatment of sequences is appropriate because it provides complete search retrieval and one need enter an extra step only in 10 percent of the cases.

No. of Steps	Percent of Entries
1 step	88.9
2 step	9.6
3 step	1.3
>3 step	0.2

Table 3. Analysis of the percentage of reactions in Theilheimer Volumes 26-35 appearing as *n*-step sequences. Ninety percent are one-step reactions.

Derwent's Journal of Synthetic Methods covers about 3,000 reactions per year including patents. The REACCS data base from this journal begins coverage from 1982 as a follow-on from Theilheimer. There will be two updates of this data base per year and the format of the data base will be essentially the same as Theilheimer.

Current Literature File (CLF) is also an ongoing file beginning in 1984 and going forward. Reactions for this file are selected by collaborating chemists in academia and industry on the basis of their expertise, but they are entered and checked by the Data Base Division of Molecular Design Limited. CLF, together with the Journal of Synthetic Methods will provide excellent coverage of the current reaction literature. CLF in particular emphasises stereochemical control in chemical reactions which is very important in much modern synthetic activity.

The cumulative number of reactions in production versions of REACCS databases as a function of time is shown in Figure 1. By the beginning of 1986 there will be 60,000 reactions, and by 1987 there is projected to be 82,000 reactions.

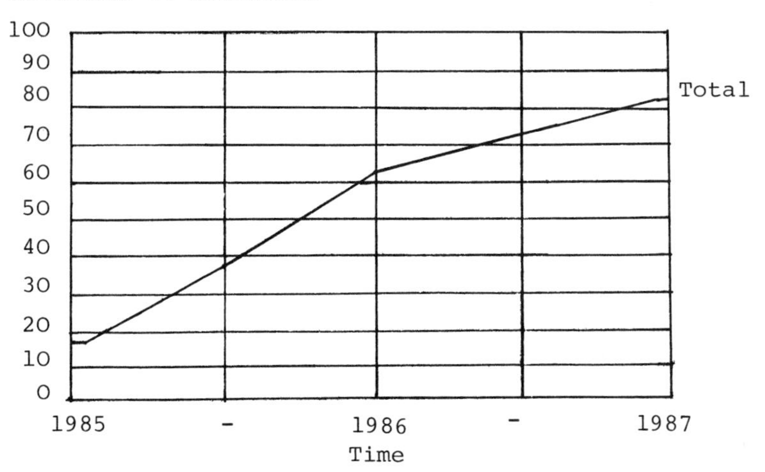

Figure 1. The projected cumulative reactions available in commercially available REACCS databases.

Fine Chemicals Directory (FCD) is a directory of 80,000 commercially available chemicals and the suppliers of each chemical. Over forty sources of intermediates and reagents are covered by this database. FCD has been available in MACCS form and has now been converted into REACCS form so the commercial availability of any compound in a reaction may be determined in REACCS. FCD will be updated annually.

The flexible data base structure of REACCS gives the data base designer complete control over the arrangement of and hierarchy of data types. A portion of the Organic Syntheses data dictionary control language for CONDITIONS is given in Figure 2. The data dictionary specifies the name of the entity, whether it is fixed or variable in length, single or multiple entries, security, data type, vocabulary control, units of measure, prompts, etc.

```
PDT=CONDITIONS
    ACCESS CODE=FM
    SECURITY=11
    LDT=SIZE
        TYPE: R
        VOCABULARY: VOLUME
        UNITS: LITERS
        INPUT PROMPT: Equipment Size:
        INPUT FORMAT: R
        OUTPUT FORMAT: R,' LITERS'
    LDT=ATMOSPHERE
        TYPE: A3
        VOCABULARY: <'HE','H2','N2','AR','AIR','CO2','CH4'>
        INPUT PROMPT: Reaction Atmosphere:
        OUTPUT FORMAT: A3
    LDT=PRESSURE
        TYPE: R,R
        VOCABULARY: PRESSURE
        UNITS: PSIG
        INPUT PROMPT: Initial Pressure:
        INPUT FORMAT: R,'-',R
        OUTPUT FORMAT: R,'-',R,' PSIG'
    LDT=TIME
        TYPE: R,R
        VOCABULARY: <0-999.99>
        UNITS: HR
        INPUT PROMPT: Reaction Time:
        INPUT FORMAT: R,'-',R
        OUTPUT FORMAT: R,'-',R,' HR'
    LDT=TEMP
        TYPE: R,R
        VOCABULARY: TEMPERATURE
        UNITS: C
        INPUT PROMPT: Reaction Temp:
        INPUT FORMAT: R,'-',R
        OUTPUT FORMAT: R,'-',R,' DEG C'
    LDT=REFLUX
        TYPE: A1
        VOCABULARY: <'Y','N'>
        INPUT PROMPT: Reflux ?
        OUTPUT FORMAT: 'Reflux? ',A
END CONDITIONS
```

Figure 2. Data structure for Organic Syntheses CONDITIONS. Shown here is a portion of the data dictionary definition language.

Molecules also have a data structure for data related to the molecule. A molecule can be involved in many reactions as reactant or product, yet will only be stored once in the data base. Similarly, a reaction is stored only once even though there may be many variations of that reaction stored. Having presented the data bases over which we will be searching, let us begin exploring reactions.

Stereospecific Electrocyclic Example. We will skip details of drawing chemical structures since that is familiar to this audience. Our first exploration is specific, 'Look for examples of opening a four-membered ring to form two double bonds in an eight-membered ring.' We first indicate we wish to BUILD a query. This provides the drawing menu. We draw the reactant shown in Figure 3 with anti-ring junctures. Ring fusion bonds are marked as CENTER, meaning they are broken in the reaction (they appear intensified on the screen). The four stereocentres are marked STEREO which is displayed by the box around the atom, meaning use the stereochemistry as a constraint on the search. We then ADDREACTANT, which adds this query structure as a reactant query.

Next we create the product query (Figure 4), generally by modifying the reactant. The two double bonds of the cyclooctadiene are specified CENTER and STEREO, meaning that the double bonds are formed in the reaction and the stereochemistry of them must be as drawn, namely, cis-trans. This structure is added as a product by the ADDPRODUCT button. The reaction query is symbolised in the upper left corner (Figure 4) by A ⟶B. By pointing to the arrow with the pointing device (light pen in this case), we see the overall reaction query shown in Figure 5.

The search is entered by pointing to SEARCH which presents the sub-menus shown in Figure 5. We wish a reaction substructure search, RSS, meaning find all reactions in which the reactant queries are substructures of the reactants and the product queries are substructures of the products. In the upper right corner, REACCS shows we are searching the Theil.60 data base containing 13,168 reactions and that we are not restricting the search with a previously specified reference list.

We view the results of the search by VIEWList, which provides a sub-menu of buttons to view the FIRST, PREVious, NEXT, or any particular ITEM on the list of hits (list 1) (see Figure 6). Our answer is reaction number 707 in volume 31 (internal registry number of 8130), by D. Bellus, et al. with the original reference. (10) Reaction conditions were 72 hours at 100° C, and the yield was 98%. The stereochemistry does indeed meet the requirements of our query. Additional substitution is allowed at all positions on the query, in this case four cyano groups are attached to the reacting centre. Further searches with modified stereochemistry of the reactant side or product side of the query gave no hits.

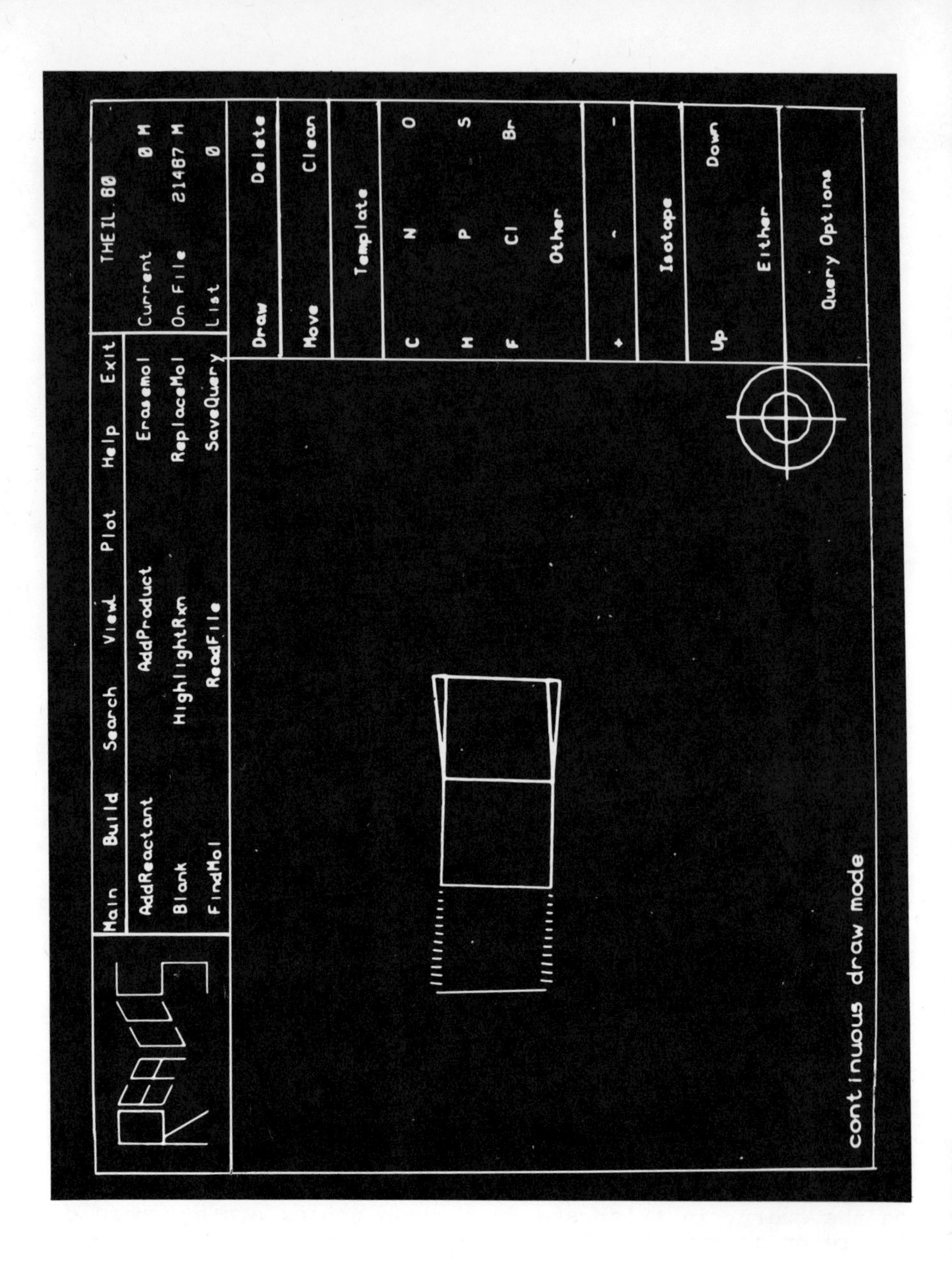

Figure 3. Reactant query drawn in BUILD mode with fusion bonds specified CENTER and stereocentres specified STEREO. ADDREACTANT adds it to the current reaction query. (Photographed from Imlac terminal.)

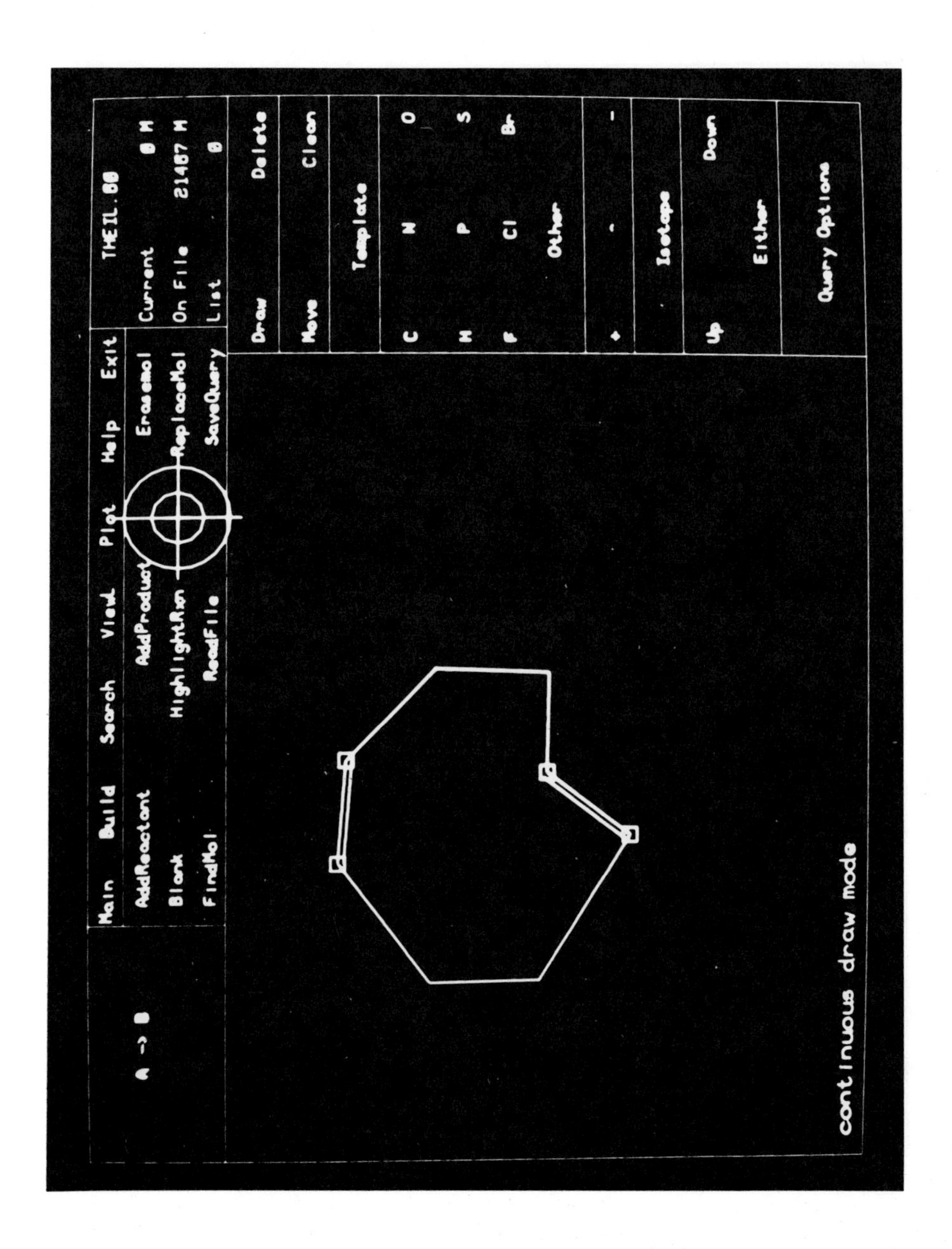

Figure 4. Product query, cis-trans-cyclooctadiene with double bonds specified CENTER and STEREO, just after ADDPRODUCT button hit.

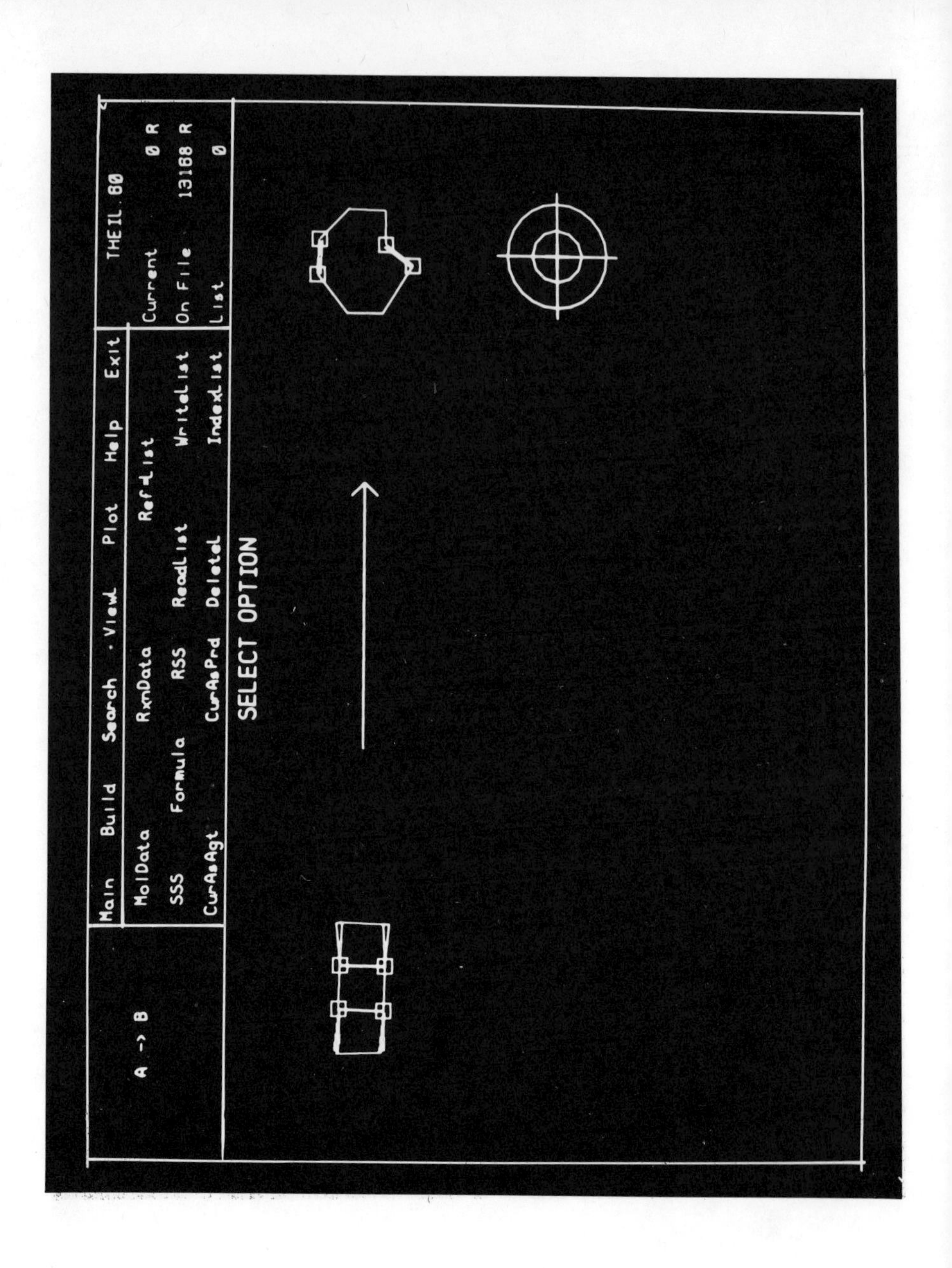

Figure 5. SEARCH mode entered with the stereospecific butane ring opening query displayed. RSS, reaction substructure search, applies this query to the 13,168 reactions in the Theil.60 data base (vols 26-35).

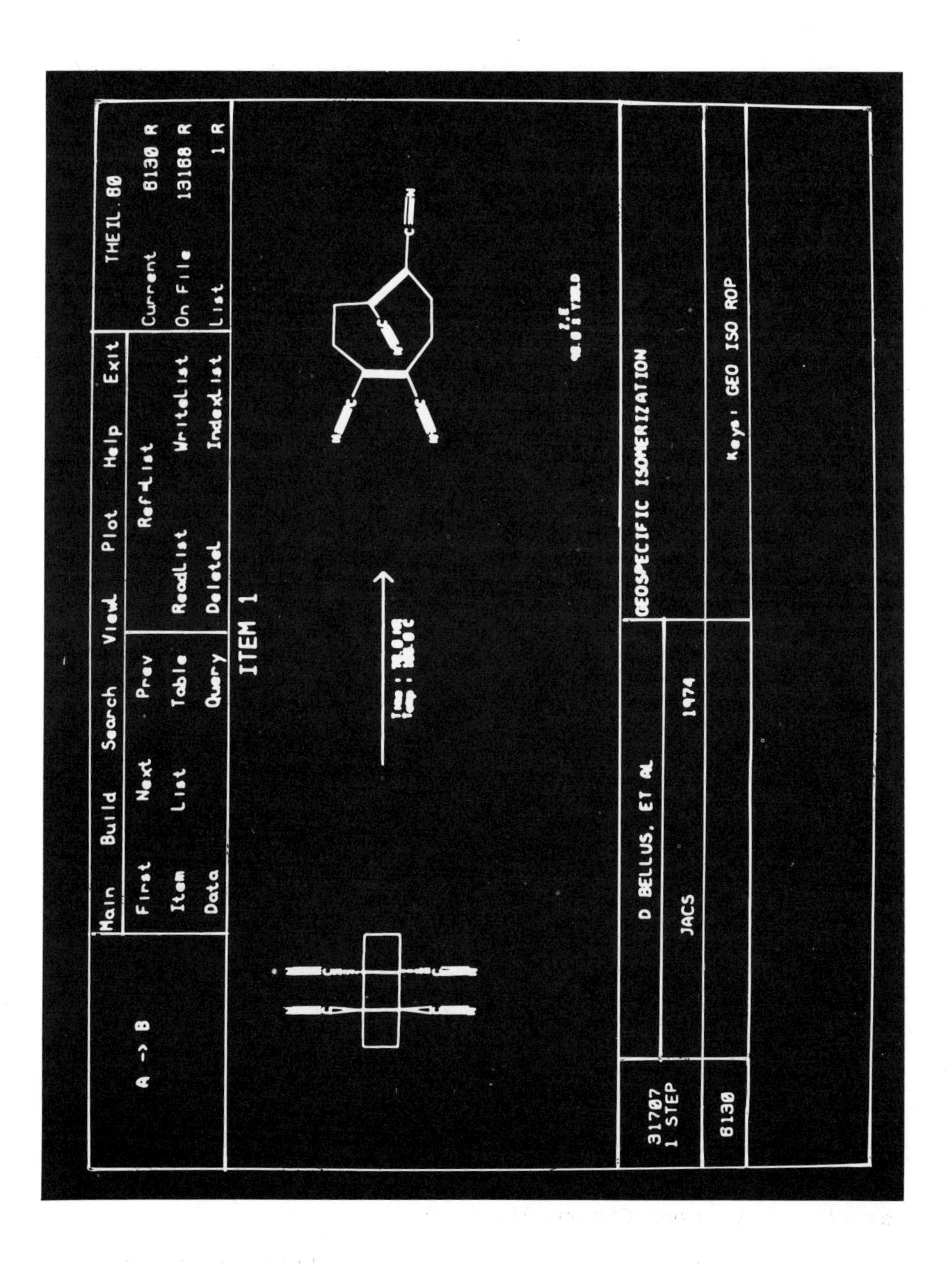

Figure 6. Viewing (VIEWL) the FIRST item on the hitlist for the RSS shown in Figure 5. Tetracycano-cis,trans-cyclooctadiene is produced in 98% yield in 72 hours at 100° C.

Browsing Reagent Uses. Let us now begin with the simple question: 'Are there any reactions where PBr_3 is used as a catalyst or as a reactant which are not simply conversion of C-OH to C-Br?' We are looking for reactions other than those with which we are familiar. The REACCS commands to answer this question are given in Figure 7.

```
Search: L0 = symbol=PBr3 as catalyst or symbol=PBr3 as reactant
[M] 1 HIT          MOL:SYMBOL = PBR3
[M] 1 HIT          MOL:SYMBOL = PBR3
---------------
[R] 10 HITS        REFERENCE LIST

Search: CSS=C-OH -> C-[Br]
** USING REFERENCE LIST
[R] 8 HITS         CSS C-OH -> C-[BR]
---------------
[R] 8 HITS         LIST 1

Search: L0 MINUS L1
[R] 10 HITS        REFERENCE LIST
[R] 8 HITS         LIST 1
---------------
[R] 2 HITS         LIST 2
Search: VIEW
CURRENT LIST: 2
View: FIRST
```

Figure 7. This query finds reactions in which PBr_3 is catalyst or reactant which are not conversions of an alcohol to a bromide.

The reference list is set to contain reactions in which PBr_3 is a reactant or catalyst, and using ORGSYN, we find ten hits. Over this list we perform a reacting centre substructure search, CSS, for reactions converting OH to Br - eight hits were found. Taking the difference we are left with two hits in List 2 that satisfy the query. We now VIEWL the FIRST hit on list 2, which is shown in Figure 8. Piperidyl benzoyl amide (0.42 moles) is treated with PBr_3 (0.43 moles) to give 1,5-dibromopentane in 65-72 percent yield plus benzonitrile. (11)

The NEXT item on hit list 2 is the reaction of diethyl ketone (0.5 moles) with bromine (1.0 moles) and a catalytic amount of PBr_3 (1 ml) to give the α,α'-dibromoketone in 72 percent yield (see Figure 9). (12)

We are alerted that the product is a potent lachrymator and readily absorbed, skin irritant - use hood. A second variation uses 100 ml of 47% PBr_3 and also gives 70-76 percent yield. The reaction is 1 of 2 steps. The second step will be the next entry in the data base, but we can find reactions of the dibromoketone easily. First we select that molecule for viewing by MOL C, resulting in the screen shown in Figure 10.

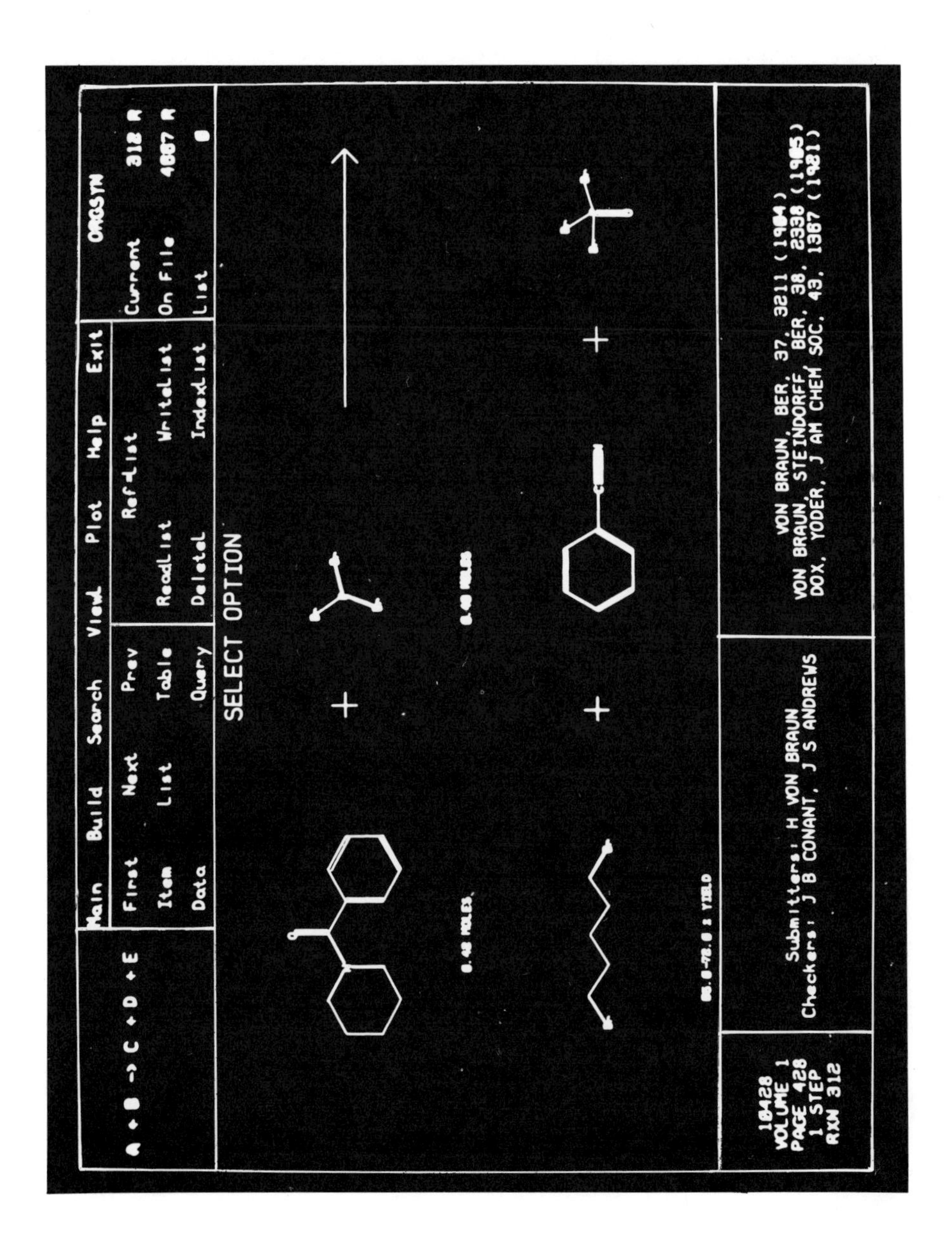

Figure 8. The first answer to query of Figure 7 from ORGSYN is quite an unusual reaction.

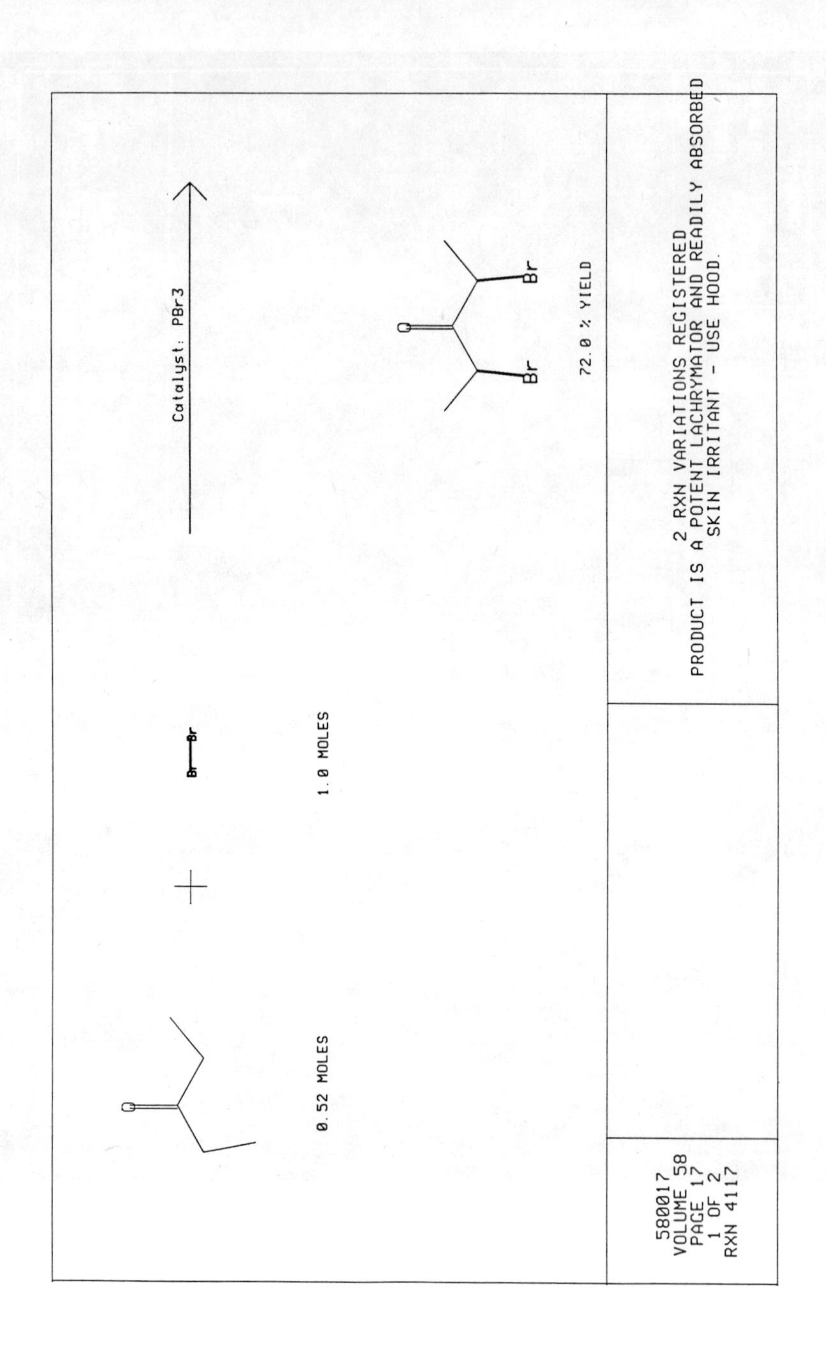

Figure 9. Halogenation alpha to a ketone is the second answer to the query of Figure 7 from ORGSYN. This figure is an example of REACCS plotting to the QMS Laserprinter.

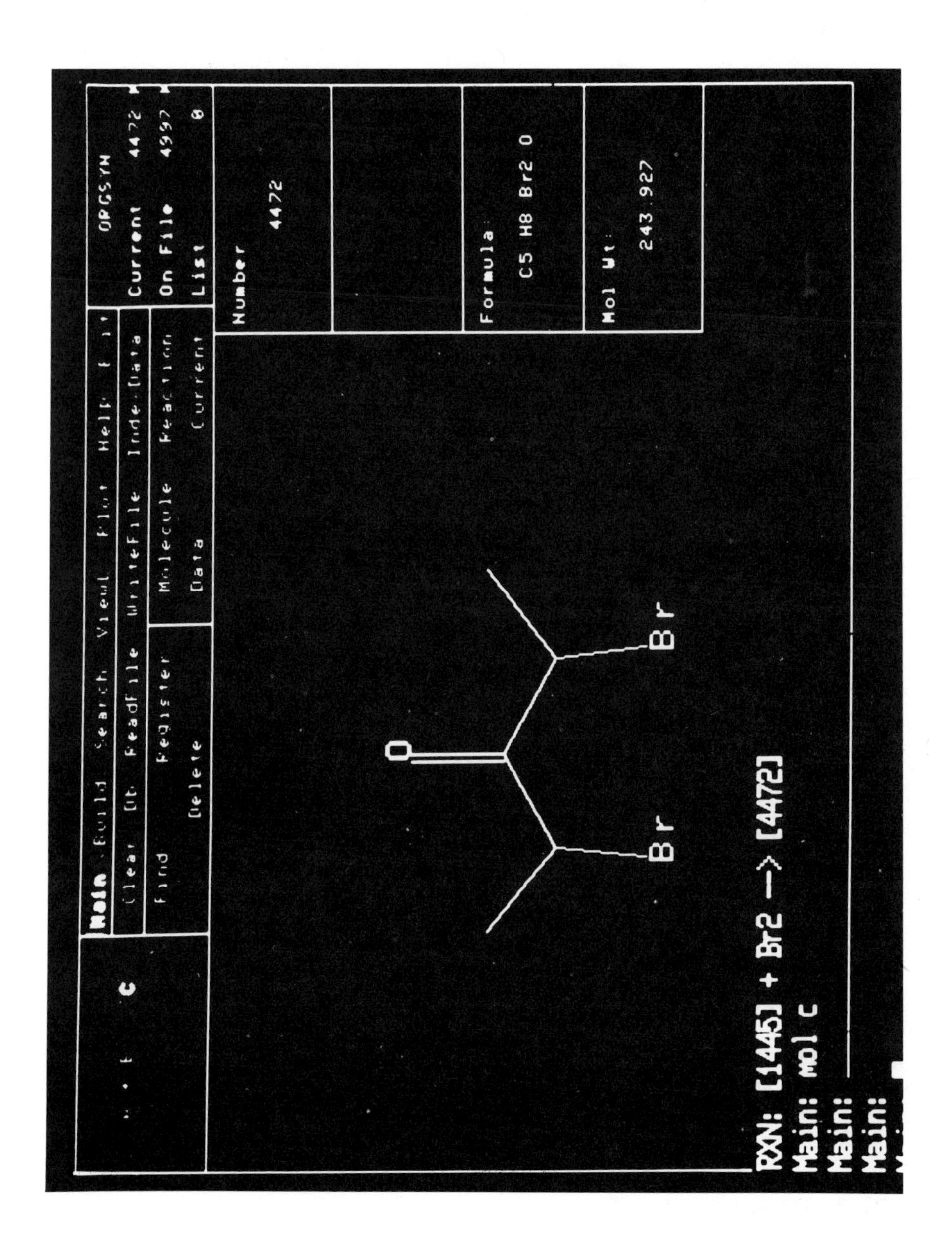

Figure 10. The command MOL C selects the product molecule from reaction 4117 in Figure 9 and displays with molecule-related data. (Photo from Envision colour terminal via Celtic camera.)

Note that the form on the screen changed to the molecule form with molecular formula and molecular weight and compound number. In the upper right of Figure 10 we see 4997 molecules on file and in Figure 9 we saw 4887 reactions on file for ORGSYN. Although there are many molecules in each reaction, there are only slightly more unique molecules than unique reactions. REACCS does not waste space storing molecules redundantly.

To find reactions in which the dibromoketone is a reactant we simply type 'Search Current as Reactant' or graphically hit SEARCH and CURASAGENT, then VIEWList 3 which contains in this case two hits, reaction 4118 and 4135, displayed in Figures 11 and 12.

Figure 11 shows a 1,4-cycloaddition to the diene furan giving a bicyclic product in 40-48 percent yield. (12) The diene is present in a four-fold excess. This reaction is the second step reported as part of the same procedure as the synthesis of the dibromodiethyl ketone (Figure 9). The second reaction (Figure 12) shows a 1,2-addition to an enamine (present in three-fold excess) giving a cyclopentenone in 64-67 percent yield. (13) We are warned that the catalyst, $Fe_2(CO)_9$, is highly toxic. This reaction was not part of the same procedure as the synthesis of the dibromodiethyl ketone but was connected by REACCS through the common chemical structure.

Wondering about how one synthesises the morpholine enamine reactant in Figure 12, we type 'MOL B' since it is the second molecule in the reaction, and that molecule becomes the <u>current molecule</u>. 'Search Current as Product' finds one hit, showing the reaction of acetophenone with morpholine in benzene with catalytic <u>p</u>-toluenesulfonic acid to give the enamine in 57-64 percent yield. (13)

Up to this point we have considered reactions of dibromodiethyl ketone itself. Now we wonder, 'What other reactions do substituted <u>bis</u>-alpha bromo ketones undergo?' Since we do not know what kind of reactions we are looking for we do not mark the C-Br bonds as CENTER bonds. Instead, we retrieve the dibromodiethyl ketone structure by FIND MOLECULE 4472 (one of many ways) so it becomes the <u>current molecule</u> and we hit SEARCH:

```
Search: sss as reactant
[R] 5 HITS     LIST 6
Search: sss as product
[R] 4 HITS     LIST 7
Search: L6 and L7
    NO HITS
Search: L6 minus L3 (L3 contains rxns in Figures 11 and 12)
[R] 3 HITS     LIST 8
Search: VIEWList
CURRENT LIST: 8
View: FIRST (Displays reaction 1 shown in Figure 13)
```

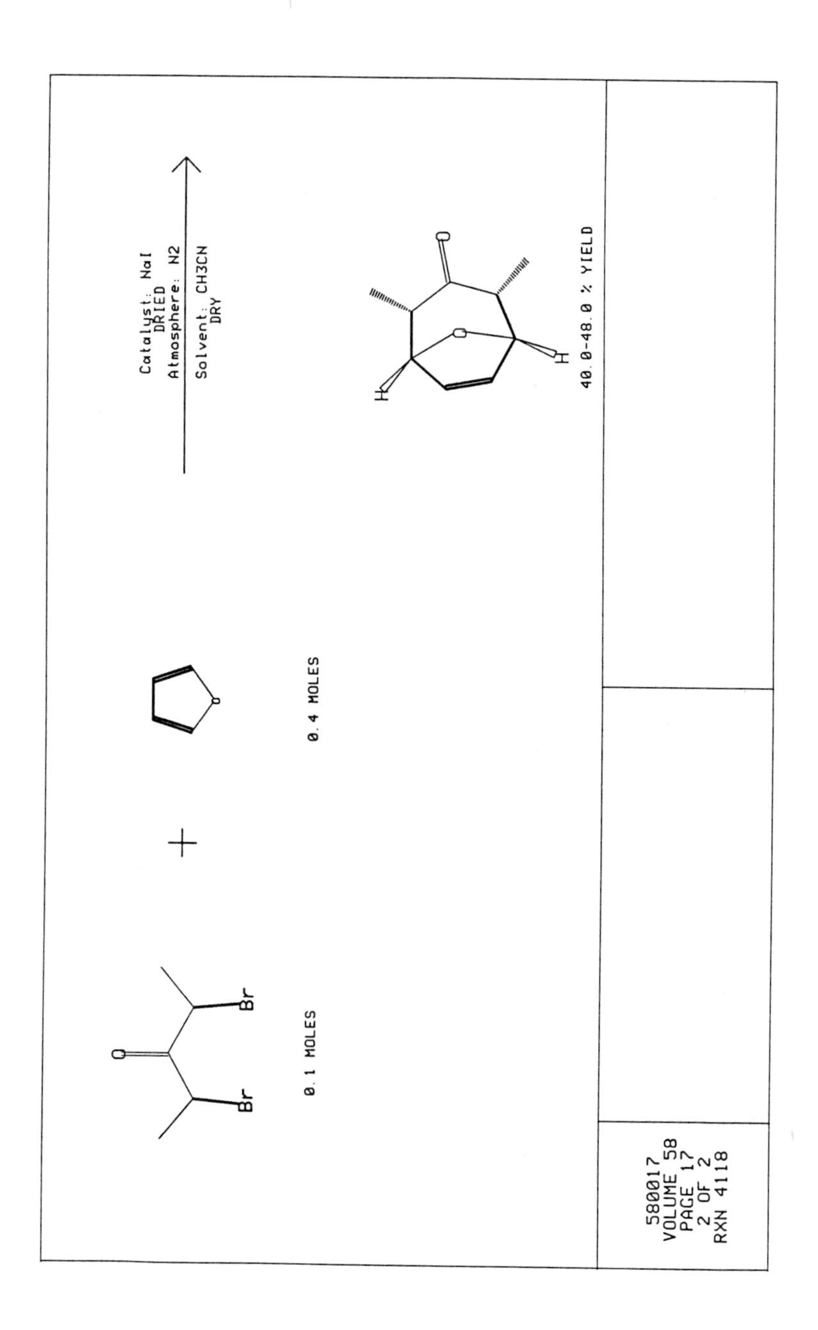

Figure 11. Asking for reactions in which dibromodiethyl ketone is a reactant found the step following the preparation of the ketone and illustrates cycloaddition with a diene.

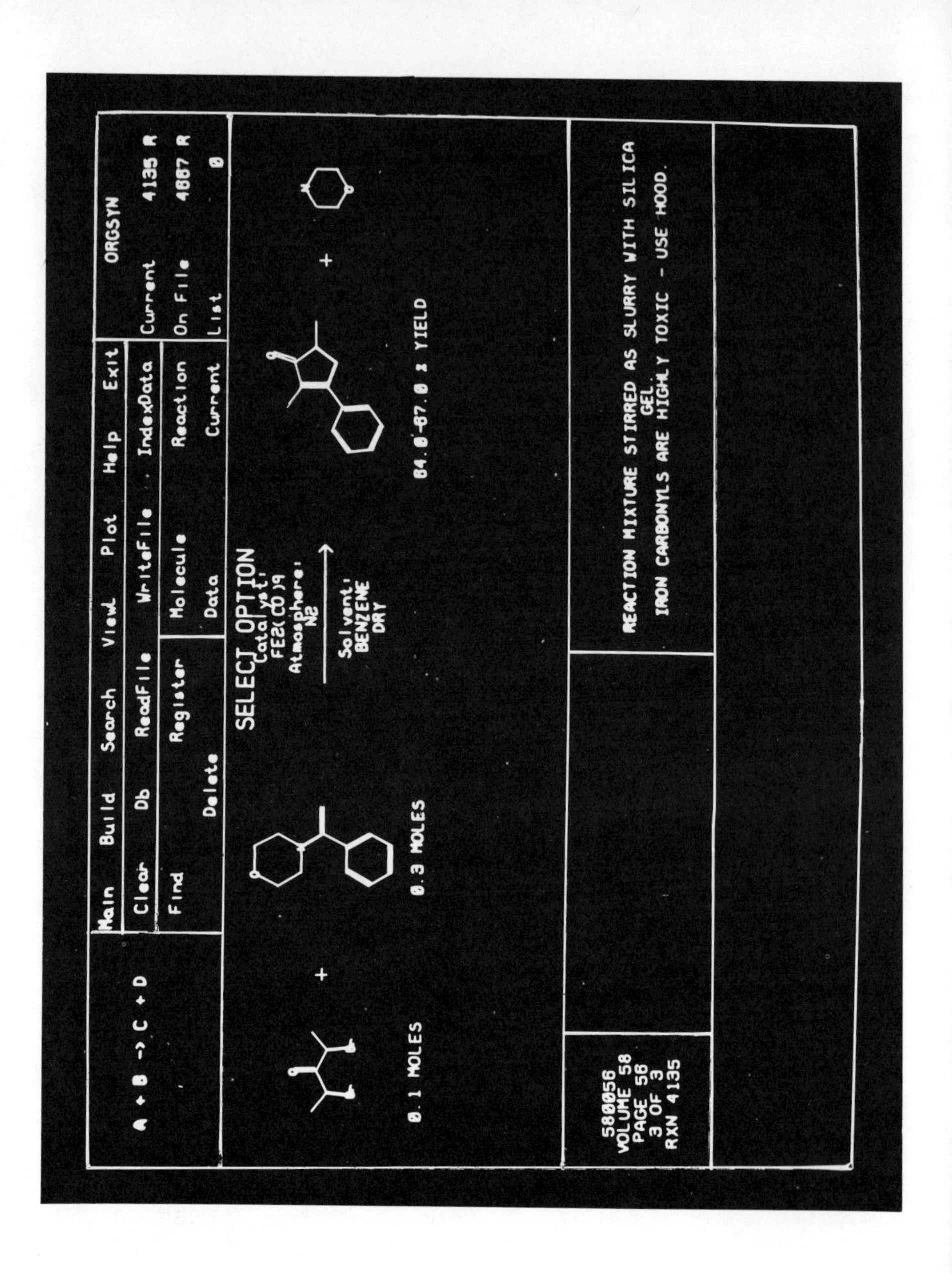

Figure 12. In the second answer to the query of Figure 7, we discover the dibromdiethyl ketone also undergoes cycloaddition with an enamine.

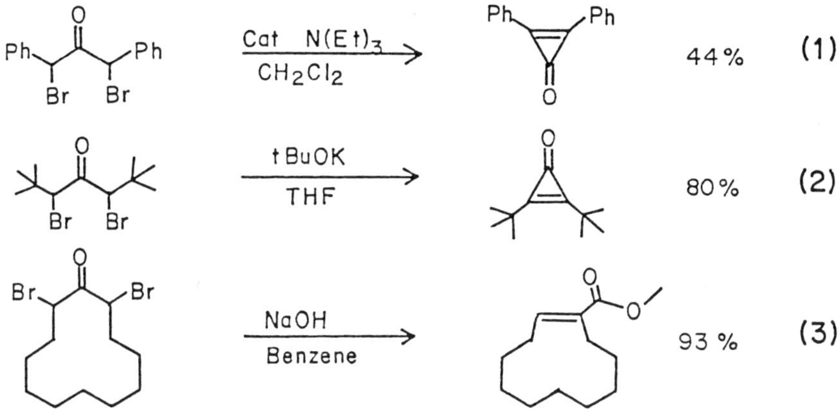

Figure 13. Allowing the dibromodiethyl ketone to be embedded as a substructure in the reactant produces three new hits from ORGSYN. The extra substitution appears to stabilise the cyclopropenone.

The fact that the intersection of list 6 and list 7 contains no hits means that none of the reactions with the dibromoketone substructure in the reactant also have the same substructure in the product, thus that substructure is being modified in the five reactions on list 6. We then subtract from list 6 the two reactions previously seen (they happen to be on list 3), and then view the three remaining reactions (Figure 13). Immediately from reactions 1 (14) and 2 (15) we discover cyclopropenes are produced when the ketone is substituted with phenyl or t-butyl groups in the alpha positions. This new information suggests that cyclopropenes may be reactive intermediates in the previous reactions. We also notice the yield for (2) of 79-83 percent is higher than that for (1) (44 percent), suggesting that the more sterically hindering t-butyl group stabilises the cyclopropenone. An important point to notice is that (1) is found in volume 5, whereas (2) is found in volume 54. REACCS has brought these physically separated reports of reactions together in time and space that might not have occurred in a manual search.

Reaction 3 (16) (Figure 13) illustrates a ring contraction which we note could occur by cyclopropenone formation, followed by attack on the unhindered ketone by methoxide.

Seeing the stable di-t-butylcyclopropenone formed in reaction 2 raised curiosity about what reactions may have been carried out with this material and we noted that this reaction was part of a reported sequence. Thus we performed a FIND REACTION 3738, the next sequential reaction to reaction 2, and found the reaction with t-butyl lithium and HBF_4

in pentane giving the tri-t-butylcyclopropene carbocation which we note is an aromatic system (see Figure 14). (15)

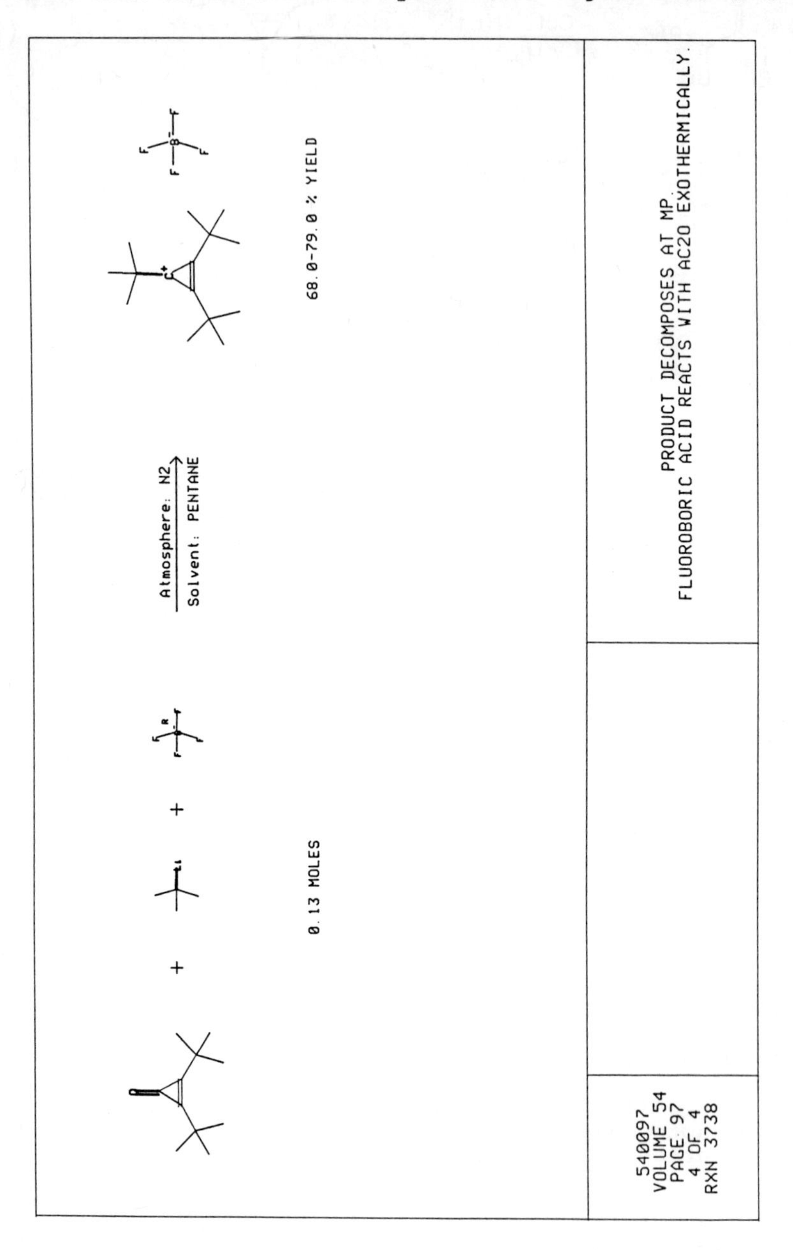

Figure 14. Di-t-butylcyclopropene can be transformed into the aromatic cyclopropenium ion.

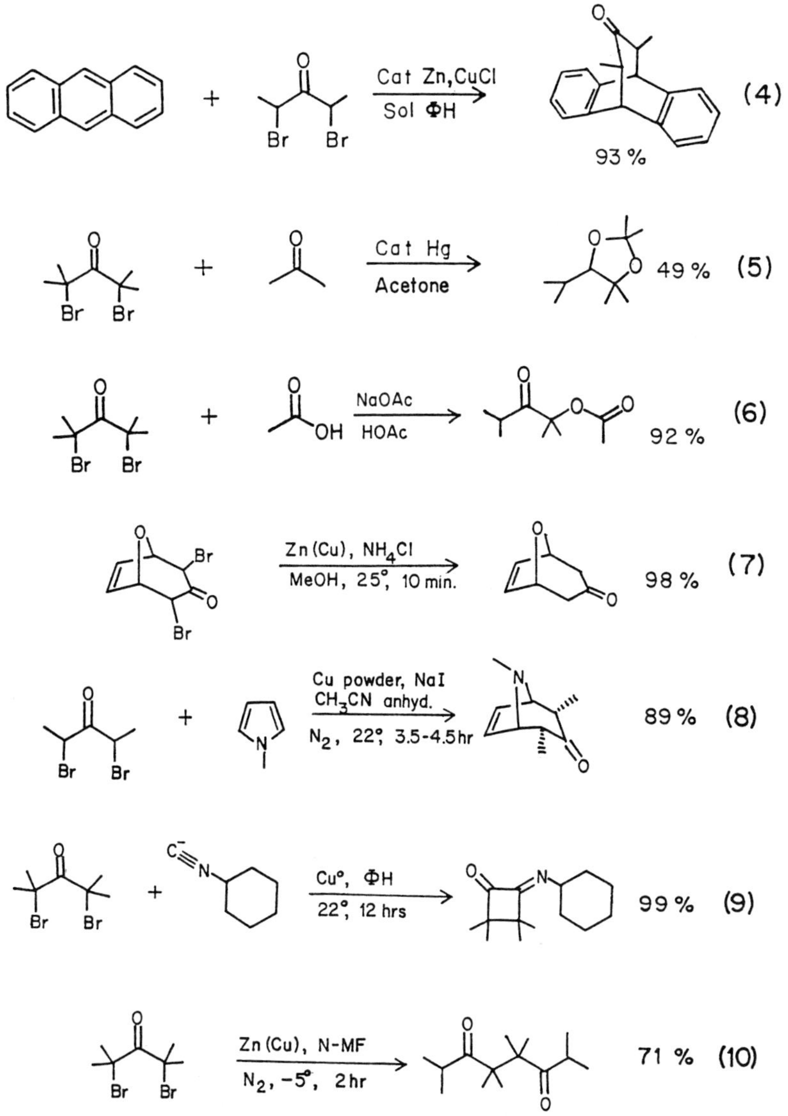

Cat Zn,CuCl; Sol ΦH — 93 % (4)

Cat Hg; Acetone — 49 % (5)

NaOAc; HOAc — 92 % (6)

Zn (Cu), NH_4Cl; MeOH, 25°, 10 min. — 98 % (7)

Cu powder, NaI; CH_3CN anhyd.; N_2, 22°, 3.5-4.5 hr — 89 % (8)

Cu°, ΦH; 22°, 12 hrs — 99 % (9)

Zn (Cu), N-MF; N_2, −5°, 2 hr — 71 % (10)

Figure 15. Searching the THEIL.60 data base of 13,168 reactions for reactions of substituted bis-alpha dibromo ketones leads to reaction types not seen in ORGSYN.

Browsing in THEIL.60. Switching from ORGSYN to another data base such as THEIL.60 is as simple as typing 'DB THEIL. 60' plus switching viewing formats to fit new data types in the new data base. Queries may be transferred between data bases. Thus with the dibromodiethyl ketone as the current query as we switch data bases, we may immediately repeat the same 'SSS as Reactant' search on the Theilheimer data base of 13,168 reactions. Doing this we obtained 13 hits, some of which are shown in Figure 15. Reaction 5 (17) represents a type of reaction not seen earlier in ORGSYN, and reaction 6 (18) appears related. We see the bromines can be removed in 98 percent yield in reaction 7. (19) Both reactions 4 (20) and 8 (21) are similar to the cycloaddition behaviour seen earlier, but reaction 9 (22) to form a four-membered ring in 99 percent yield is yet another new type of cycloaddition not seen in ORGSYN. Finally, with copper, we see coupling of the bromide to form a dimer plus reduction (reaction 10). (23)

CONCLUSION

Through this excursion we have seen the nature of some of the commercially available and supported data bases for REACCS, which together by 1987 will comprise 82,000 reactions. All these reactions are being entered by Molecular Design Limited under rigorous quality control standards. The Current Literature File (CLF) and the Journal of Synthetic Methods (JSM) files will cover the most recent literature and will be ongoing. Together these data bases will contain a large proportion of the chemically interesting reactions, all with stereochemistry, and quantitative supporting data including original literature references. These reactions are the factual observations of chemical experiments, not interpretations thereof.

Through our first example, we demonstrated how simple it is to prepare even a complicated stereochemical search question, and how accurately and specifically REACCS finds answers to such a question. In our second example we demonstrated how one can look for unusual reactions across multiple data bases.

We have shown that, contrary to popular opinion, it is possible to browse computerised reaction data bases to learn unexpected information. We began by exploring reactions of PBr_3 other than conversion of alcohols to bromides. Seeing that one such reaction is the preparation of bis-alpha dibromo ketones, we wandered into the chemistry of these dibromo ketones and discovered a wide variety of types of chemical reactions, including in sterically hindered cases the intermediacy of cyclopropenones. Along the way we explored the synthesis of some of the reactants and the chemistry of some of the products, including the formation of the aromatic cyclopropenium cation. We easily linked one reaction to another to follow sequences of reactions, and combined hit lists of reactions with logic to exclude unwanted features or reactions. The total time required for this excursion was a couple of hours, most of which was spent by the chemist thinking and analysing results. Search times were a matter of seconds.

By combining searches over ORGSYN and THEIL.60, one benefits by the small collection of very high quality data in ORGSYN plus the larger diversity of reaction types offered by the Theilheimer data base. Our studies of journals and authors covered in Theilheimer indicated that it does have great breadth of coverage. Although not presented in this paper, we could have transferred any of the molecules from any reaction to the Fine Chemicals Directory data base and with REACCS discovered the suppliers for the material if it is commercially available. Finally we found the quantitative data in these data bases very useful for analysing the reactions, determining which compound is the limiting one, and comparing yields.

One very clear conclusion of this paper is that computerised reaction data bases when used together with the powerful retrieval tool, REACCS, permit the old and the new reactions to appear adjacent in time and space. Multi-volume works appear as a single unified entity. The consequence of this to the chemist is that it increases the opportunity to analyse these once scattered experimental observations and see new relationships among them, new analogies, new discoveries.

ACKNOWLEDGEMENT

One of us (WTW) did the browsing and writing, all participated in building REACCS. TM and DG brought REACCS to version 6.0 and thank Jim Nourse for helpful discussions in that regard. We also thank Mike Weinshelbaum for his work in building the data bases and conducting the statistical studies of THEIL.60. We are grateful to Lisa Helfend and Cathy Calderwood for photographic assistance.

REFERENCES

(1) Wipke, W. T., 'Exploring Reactions with REACCS'. Presented at the 188th National Meeting of the American Chemical Society, Philadelphia, PA, August 1984.

(2) Valls, J. 'Reaction Documentation'. In Computer Representation and Manipulation of Chemical Information; Wipke, W. T.; Heller, S. R.; Feldmann, R. J.; Hyde, E., (eds.) John Wiley: New York, 1974, pp 83-104.

(3) Wipke, W. T.; Dill, J. D.; Peacock, S.; Hounsell, D., 'Search and Retrieval Using an Automated Molecular Access System', Presented at the 182nd National Meeting of the American Chemical Society, New York, NY, August

1981.

(4) 'How to Synthesise Molecules - on a CRT', Chemical Week, October 14, 1981, 73-74.

(5) 'Software Retrieves Reaction Data'. Industrial Chemical News, April 1982, 3(4).

(6) 'Computer System Searches Chemical Reactions'. Chem. Eng. News, April 12 1982, 60(15), 92.

(7) Marson, S. A.; Peacock, S. C.; Dill, J. D.; Wipke, W. T., 'Computer-Aided Design of Organic Molecules', Presented at the 177th National Meeting of the American Chemical Society, Honolulu, Hawaii, April, 1979.

(8) Cruickshank, P. 'How a Computer Program Helped Build a Chemical Registry'. Industrial Chemical News, April 1984, 5(4), 36-37.

(9) Adamson, G. W., Bird, J. M., Palmer, G., Warr, W. A. 'Use of MACCS within ICI'. J. Chem. Inf. Comput. Sci., 1985, 25(2), 90-92.

10) Theilheimer 31, 367; Bellus, D., et al., J. Amer. Chem. Soc., 1974, 96, 5007; Trost, B. M.; Herdle, W. B., J. Amer. Chem. Soc., 1976, 98, 1988.

(11) von Braun, H., Org. Syn., 1, 428 (Checker: Conant, J.B.; Andrews, J. S.); von Braun, Ber., 37, 3211 (1904); von Braun, Steindorff, Ber., 38, 2338 (1905); Dox, Yoder, J. Amer. Chem. Soc., 43, 1367 (1921).

(12) Ashcroft, M. R.; Hoffmann, H. M. R., Org. Syn., 58, 17; Fierz, G.; Chidgey, R.; Hoffmann, H. M. R. Angew. Chem. Int. Ed. Engl., 13, 410 (1974).

(13) Noyori, R.; Yokoyama, K.; Hayakawa, Y., Org. Syn., 58, 56 (Checker: Haire, M. J.; Sheppard, W. A.); Noyori, R.; Yokoyama, K.; Makino, S.; Hayakawa, Y., J. Amer. Chem. Soc., 94, 1772 (1972).

(14) Org. Syn., 5, 514 (Checker: Lokensgard, D. M.; Chapman, O. L.); Breslow, R.; Posner, J.; Krebs, A., J. Amer. Chem. Soc., 85, 234, (1963).

(15) Ciabattoni, J.; Nathan, E. C.; Feiring, A. E.; Kocienski, P. J., Org. Syn., 54, 97 (Checker: Leichter, L. M.; Masamune, S.).

(16) Wohllebe, J.; Garbisch, Jr., E. W., Org. Syn., 56, 107 (Checker: Diakur, J. M.; Masamune. S.)

(17) Theilheimer, 35, 491; Fry, A. J.; Ginsburg, G. S.; Parente, R. A., J. Chem. Soc. Chem. Comm., 1978, 1040-1041.

(18) Theilheimer, 32, 92; Fry, A. J.; O'Dea, J. J., J. Org. Chem., 1975, 40, 3625.

(19) Theilheimer, 31, 43; Noyori, R. et al., J. Org. Chem., 1975, 40, 806.

(20) Theilheimer, 35, 490; Giguere, R. J.; Rawson, D. I.; Hoffmann, H. M. R., Synthesis, 1978, 902-905.

(21) Theilheimer, 31, 420; Fierz, G.; Chidgey, R.; Hoffmann, H. M. R., Ang. Chem., 1974, 86, 444.

(22) Theilheimer, 30, 457; Ito, Y., et al., Synth. Commun., 1974, 4, 87.

(23) Theilheimer, 29, 393; Chassin, C.; Schmidt, E. A.; Hoffmann, H. M. R., J. Amer. Chem. Soc., 1974, 96, 606.

10 SYNthesis LIBrary

D. F. Chodosh
Smith, Kline and French Laboratories

The introduction of minicomputer and microcomputer technology has, in one short decade's time, remarkably influenced the chemical research laboratory. Researchers in analytical and physical chemistry have, of necessity, gained 'computer literacy', adapting to the technology as sophisticated laboratory automation and computer-based instrumentation became widely available for their laboratories. Until recently, with the exception of efforts toward retrosynthetic synthesis design and despite their sizable proportion of the chemical science community, slow progress has been made toward development of computer technologies for the synthesis chemist (1-4). It is therefore not surprising that, as a group, synthesis chemists remain more aloof to this technology than their physical and analytical colleagues.

The general utility of computer systems for the management of collections of information is well demonstrated. The 1970s saw the development of time share services that permitted ready access to bibliographic chemical databases using keyword, author and citation searching. More recently computer algorithms have been developed to permit storage and retrieval of chemical structures and substructures. These structurally oriented databases may be interrogated by using a suitable graphical method to construct a structural question with which to query a database. These latter innovations led directly to the development of 'registry systems' for the management of public domain (e.g., CAS ONLINE) (5) and proprietary (e.g., MACCS, COUSIN) (6-7) chemical structure inventories, allowing chemists to seek structural answers to structural questions. Compound registry systems however remain a step away from the synthesis aspect of chemical research. To achieve this next step we have developed SYNthesis LIBrary (SYNLIB tm) a knowledge based expert system for chemical reactions (8).

SYNLIB has been designed to communicate chemical structures and chemical reaction concepts with synthesis chemists in their pseudo-natural language: chemical structures. This sophisticated interface permits chemists to communicate chemically meaningful concepts to the computing system in much the same fashion as chemists communicate among themselves. This approach, though computationally complex, reduces barriers to the use of the system: chemists no longer must learn an artificial computer language with which to describe their artificial chemical language. The SYNLIB search algorithms function transparently to the end user chemist taking chemically complex questions and seeking solutions from a collection of synthesis knowledge. SYNLIB search methods have been designed to manipulate knowledge bases of chemical reactions cited from the published literature. In contrast with retrosynthetic predictive systems that rely on sets of synthesis rules ('transforms') to extrapolate chemical synthesis, SYNLIB presents factual synthesis information. The chemist supplies the 'synthesis art' during the SYNLIB worksession while SYNLIB provides the flexibility of dynamic search and query modification to permit rapid exploration of synthesis concepts and nuances. Through this interactive machine-chemist approach synthesis statements (i.e. facts) can be more easily evaluated in context, the system serving to assist the chemist in formulating and evaluating synthesis concepts.

SYNLIB has been developed through a collaborative research effort among scientists at Smith Kline & French Laboratories and Professor W. Clark Still and colleagues at Columbia University, NYC, NY. Concurrent with the software development an international consortium of synthesis chemists has been established (32 universities, at present) to assist in the evaluation of the chemical literature.

The utility of SYNLIB is linked to the breadth and depth of its associated chemical reaction knowledge base. The knowledge base was conceived, a priori, as an evaluated collection of representative chemical reactions as opposed to an exhaustive accumulation of published reaction chemistry: knowledge, not data. Knowledge base development therefore requires the active participation of synthesis chemists to select and abstract 'important' published chemistry and to augment literature citations with scientific evaluations. The academic community has been approached and has enthusiastically responded to this endeavour. Members of the worldwide consortium evaluate current published literature, historical literature collections and contribute to the knowledge base from their specialised areas of expertise. In return for their participation the university researchers receive current versions of the software and the database. SYNLIB performance and utility has steadily increased in parallel to the growth of the knowledge base (12,000 reactions March 1984; 21,600 reactions March 1985) and will continue to improve through the addition of new searching methods and through growth of the reaction library. In addition to the master knowledge base users may create, maintain and search any number of individual proprietary or specialised collections.

Throughout the development effort synthesis chemists have participated in the design and evaluation of the emerging system. Three key design principles were demanded of the development team:

1) Easy to learn

Chemists communicate with SYNLIB in their natural language: chemical structures. The graphical drawing procedure resembles the sketching of structures with pen and paper. Complex chemical diagrams are quickly constructed and modified with a minimum of drawing instructions. The chemist controls SYNLIB by selecting options displayed on the CRT screen ('lightbuttons').

2) Inexpensive terminals and communications

To encourage the proliferation of terminals into the laboratories SYNLIB is designed for use with low cost, low resolution video terminals. This decentralisation allows hands-on access to computer technology by the laboratory scientist. The graphical method has been developed for use at low-to-moderate host-terminal communication speeds (MODEM) to minimise cost and maximise access. Terminals in use include Macintosh, IBM-PC, VT640, VT650, Envision, AED, Lundy and Tektronix 4010/4113 CRT devices with lightpens, mouses and/or datatablets.

3) Hands-on access by laboratory scientists

By directly interfacing with the computer system scientists can rapidly research information and dynamically explore chemical nuances. This hands-on approach eliminates any possible reinterpretations of complex synthesis questions by intermediaries. This dynamic interaction of the scientist with the computing system encourages the formulation of new questions and the exploration of new ideas.

SYNthesis LIBrary search of chemical reaction collections is described below.

1) INPUT MODE

Chemical structure diagrams are constructed in INPUT MODE (through the use of a lightpen, datatablet or mouse) in conventional structural format. A 'template' file of predrawn structures may be used to facilitate structure drawing. In addition to a master template file available to all users, each chemist may construct a local template file for personal use. Templates are retrieved by user designated names.

By selecting the DRAW function in INPUT MODE new atoms and bonds may be created in the drawing box. Atom types are modified (oxygen, nitrogen ...) by selecting the appropriate 'lightbutton' followed by selecting the appropriate atom(s) in the drawing box. Atom types (elements) not shown as individual lightbuttons (C, O, N, S, X=halogen) may be designated as element Z (Z1, Z2, Z3) and subsequently defined by the chemist thereby providing access to the entire periodic table.

The last step in the drawing of a TARGET structure is the selection of the atoms that must structurally match in the subsequent search procedure. The SUBS (SUBStructure) lightbutton is used to designate these atoms which are labelled with asterisks in the drawing box.

The evaluation of the 'reactive atoms' plays a key role in SYNLIB searching. The entry of new reactions to the knowledge base requires the chemist to designate the 'reactive site' - not only those atoms and bonds that are modified during the chemical synthesis (a procedure that could well be automated) but also the neighbouring atoms influencing or directing the reaction chemistry. This interpretation of reactivity and neighbouring group effects captures interpretations of the synthesis chemist. Through this approach SYNLIB allows variable search lattitudes permitting users to search for generic or precise matches to the TARGET structure.

2) SEARCH SELECT MODE

SEARCH SELECT MODE permits searching for:

a) generic structural matches to the TARGET (BROAD SEARCH)
b) precise structural matches to the TARGET (NARROW SEARCH)
c) recall of a specific entry
d) search by journal reference (journal name and/or page and/or year

3a) SEARCH MODE: BROAD SEARCH

The BROAD SEARCH procedure requires:

$$T^* \xrightarrow[\text{matching}]{\text{forward}} P^*$$

that is, all of the substructured atoms in the TARGET must be matched to substructured atoms in the reaction PRODUCT. All other non-structural constraints that may have been specified must also match (e.g., yield, #steps, reaction conditions, ...) - see CONSTRAINT MODE.

The BROAD SEARCH function allows chemists to retrieve entries in which the highlighted portion of the TARGET (TARGET substructure atoms) are found in the highlighted portion of the PRODUCT (reacting site and influencing neighbouring atoms).

Upon initiation of the search the first matching entry is retrieved and displayed. A 'look ahead' search allows SYNLIB to continue searching while the displayed entry is examined.

BROAD SEARCH parameters that are available include:

- COUNT may be selected to complete the search of the knowledge base, assemble a match list and report out the number of entries awaiting display; the chemist may elect to return to INPUT MODE to respecify the TARGET if too many or too few matches are identified.

- OMIT may be selected to calculate a TRANSFORMATION STATISTIC of the displayed reaction. This value statistically describes bond formation/breakage and atom addition/deletion. Subsequent searching will omit the display of 'similar' reaction chemistry.

- FOCUS calculates and uses the TRANSFORMATION STATISTIC to subsequently display only 'similar' reactions during continued searching. FOCUS and OMIT may be applied to a COUNTed match list to narrow the list to include or exclude specific types of reaction chemistry.

  The FOCUS and OMIT functions are particularly powerful utilities that permit the chemist to initiate a search with a highly generic TARGET (six or seven atoms) and begin browsing the knowledge base. Upon encountering an interesting reaction SYNLIB may be FOCUSed on that reaction to concentrate the search on a particular type of bond or atom change.

- RECOVR recovers FOCUS and OMIT requirements from an interactive search procedure or from a COUNTed match list. Searches may then be continued with the chemist optionally FOCUSing or OMITing reactions as encountered.

- SRCH continues the search procedure to display the next matching entry from the knowledge base. The search may be interrupted (RTRN) to return to INPUT MODE for respecification of the TARGET. In this manner the chemist may quickly determine if the TARGET is too specific or too generic or may adjust the TARGET and re-initiate searching.

Other modes (TTY, REACTION, CONSTRAINT) may be entered by appropriate lightbutton selection.

3b) SEARCH MODE: NARROW SEARCH

The NARROW SEARCH function requires that the BROAD match criteria be met along with the condition that all substructure atoms in the product of the entry be found (but not necessarily substructured) in the TARGET chemical structure. Restated, the match requires that all chemically important atom centres be represented in the TARGET.

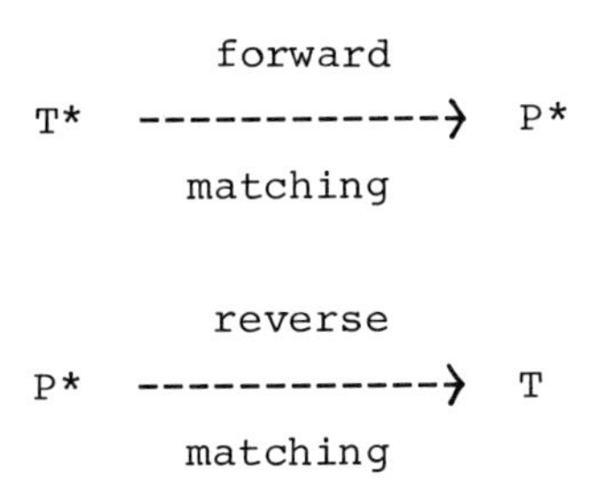

All other non-structural constraints that may have been selected in CONSTRAINT MODE must also match the entry for a successful retrieval. The two step NARROW matching algorithm allows chemists to retrieve reactions that are more precisely matched to the TARGET structural question and which therefore will describe more closely applicable reaction chemistry.

4) CONSTRAINT MODE

CONSTRAINT MODE is used to specify boundaries (constraints) to be used, along with the TARGET, in searching procedures. The mode may be entered from either INPUT MODE or SEARCH MODE thereby allowing interruption of an ongoing search to quickly specify or modify constraints.

The entire information content of each knowledge base entry may be examined and used to constrain the matching procedure: reaction conditions, atom additions/deletions, yield, synthesis steps, ... In CONSTRAINT MODE the chemist selects the appropriate matching requirements that will narrow the retrieval to entries more closely matched in context to the synthesis problem at hand.

The CONSTRAINT MODE provides the chemist with easy-to-use methods for rapidly restricting searches to relevant chemistry. With the judicious selection of constraints SYNLIB permits effective searching with highly generic TARGET structures. By way of example, one could quickly restrict a search for glycol syntheses to reactions which add oxygen while breaking C=C in yields greater than 70%. These methods allow synthetically meaningful search approaches with fairly simple TARGET structures.

CONSTRAINT MODE features include requiring/forbidding

reaction conditions, demanding specific types of bond formation or breakage, demanding specific types of atom additions or deletions, demanding a specific bond be formed during the synthesis procedure (STRATEGIC BOND), cis/trans olefin specificity, aromaticity perception, publication date requirements, limiting the number of synthesis steps and minimum yield requirements. The search may be extended to the complete product structure rather than restricting structural searching to the 'reactive substructure': this is a particularly useful feature for searching functional group transformations as in Figures 11 to 13.

5) REACTION MODE

REACTION MODE allows:

a) entry of new reactions to the knowledge base
b) edit of knowledge base entries
c) specification of a REQUIRED REACTION for matching in subsequent searches.

To facilitate the entry of new reactions to the knowledge base, REACTION MODE displays the TARGET chemical structure (from INPUT MODE) on both sides of the reaction arrow. The starting material side structure is then modified to the appropriate chemical structure using the identical light-button functions as in INPUT MODE: see Figures 14 to 17.

Once the structural parameters have been input, the reaction entry is annotated with text to appropriately describe the reaction: conditions and interfering functionality (keywords from standard SYNLIB tables), literature reference, yield, synthesis steps, reagents, solvents, any free format text comments from the evaluating chemist. The annotated text is parsed during the entry procedure to set bit maps. This procedure permits rapid searching on the basis of reaction conditions, steps, yield, reference, etc. The entire text annotation is searchable (CONSTRAINT MODE) for required and/or forbidden text characters (strings).

This virtually rule-free data format approach was adopted to relieve the chemist from arduous and restrictive 'data registration' thereby allowing the chemist to focus on the chemistry while de-emphasising the computational system. The SYNLIB searching procedures coupled with CONSTRAINT MODE and free format text searching capabilities more than compensate for the relaxation of rigorous data fields and definitions.

In addition to the entry and editing of knowledge base entries REACTION MODE allows the chemist to construct a chemical reaction question and to constrain subsequent searching by the TRANSFORMATION STATISTIC of this reaction (an a priori FOCUS). The REQUIRED REACTION function may be selected to demand either an exact statistical match of the atom and bond changes in the depicted chemical reaction or may be used to restrict the retrieval to reaction entries which at least accomplish these modifications but may further modify atoms or bonds. Coupled with TARGET structure matching the REQUIRED REACTION function permits chemists to conduct highly directed and efficient searching using generic TARGET structures: see Figure 18.

The REQUIRED REACTION feature complements the FOCUS function which permits the chemist to restrict searching only upon encountering the requisite reaction during the interactive search.

6) TTY MODE

This mode provides TELETYPE question/answer dialogue for various utility functions including saving and recalling structures/reactions from disc files (porting structures between assorted computer programs), and examination of the internal data structures of individual entries. The most useful TTY function for the typical end-user chemist is the PLOT function which globally plots the TARGET, CONSTRAINTS, any REQUIRED REACTION and all matches of the question from the knowledge base. Any individual SYNLIB CRT screen may be PLOTted by selecting the appropriate PLOT lightbutton during the worksession. The plotted screens are automatically assembled into one plotter output file, condensed 8/page (or optionally, 2/page), and are available for plotting on a central plotter device at the conclusion of the worksession. Versetec V-80, Zeta, and lasergraphics devices are currently in use for SYNLIB plot report generation. Implementation of other plotter devices is quite straightforward.

An example of the output produced by SYNLIB in a BROAD SEARCH is shown below, in Figure 19.

The SYNthesis LIBrary system continues to evolve in response to the needs and recommendations of the large numbers of synthesis chemists now using the system. A new release of the software is planned for late 1985. The incorporation of new 'screens' for rapid structure matching and modification of the data structures has permitted a doubling of search speeds; substructure searching will be extended to include both starting material and product chemical structures thereby allowing full exploration of both formation and reaction chemistry in addition to providing capabilities for yet more precise searching of specific chemical transformation. A word of caution though must be raised. While SYNLIB can and does serve well as both a compound and chemical reaction registry system the value of the system to the synthesis chemist end-user is remarkably diminished if overly specific searching is employed. Our experiences indicate that the true power of the SYNLIB system lies in the ability to interactively and dynamically explore, modify and re-explore the knowledge base, typically using generic structural TARGET questions augmented with judicious constraint requirements which in turn are derived from synthesis concepts. The chemist at SK&F is referred to CAS ONLINE or other registry systems if location of a particular chemical entity is the intended goal. The basic design of the SYNLIB system, while permitting exact structural and non-structural data matching, encourages a more generic search approach which has proven especially effective for the manipulation of chemical reaction information and for the communication of chemical synthesis concepts.

Acknowledgements

The contributions of Professor W. C. Still and colleagues (Columbia University), Dr Wilford Mendelson and Mr Joseph Durkin (SK&F) and members of the SYNLIB consortium are gratefully acknowledged.

REFERENCES

1. Gelernter, H.L., Sanders, A. F., Larsen, D. L., Agarwal, K. K., Bovie, R. H., Spritzer, G. A. and Searleman, J.E., Science, 1977, 197, 1041.

2. Corey, E. J. and Wipke, W. T., Science, 1969, 166, 178.

3. Corey, E. J., Johnson, A. P. and Long, A. K., J. Org. Chem., 1980, 45, 2051.

4. Long, A. K., Rubinstein, S. D. and Joncas, L. J., Chem. and Eng. News, 1983, 61, 22.

5. Dittmar, P. G., Stobaugh, R. E. and Watson, C. E., J. Chem. Inf. Comp. Sci, 1976, 16, 111; and references therein.

6. Wipke, W. T., Dill, J. D., Peacock, S. and Hounshell, D., 'Search and Retrieval Using an Automated Molecular Access System', 182nd National Meeting of the Amer. Chem. Soc., New York, Aug. 1981.

7. Hagadone, T. R. and Howe, W. J., J. Chem. Inf. Comp. Sci., 1982, 22, 182; Howe, W. J., Drug Inf. J., 1984, 18, 185.

8. Chodosh, D. F. and Mendelson, W. L., Pharm. Tech., 1983, 7, 90; Chodosh, D. F. and Mendelson, W. L., Drug Inf. J., 1983, 17, 231; Chodosh, D. F., 'Applications of Chemical Graphics at the Bench', Proceedings of Short Course on Hardware and Software Solutions to the Laboratory Data Management Problem, Pittsburgh Conference, Atlantic City, NJ, March 1984.

SELECT TEMPLATE - MASTER FILE

DESCRIPTION	NAME
ACYCLIC HYDROCARBONS	A5 - A15
SATURATED MONOCYCLES	3 - 16
UNSATURATED MONOCYCLES	3U - 10U
BYCYCLIC RINGS	B110 - B333
Tetralin (tetrahydronaphthalene)	TLIN
Naphthalene	NAPH
Anthracene	ANTH
Perhyroanthracene	PANT
Phenanthrene	PHEN
Perhydrophenanthrene	PPHE
D-Glucose (pyranose form)	GLUC
Progesterone	PROG
Testosterone	TEST
Cortisone	CORT
Estradiol	ESTR
D-Ribose (furanose form)	RIBO
Monounsaturated Prostaglandin	PG1
Diunsaturated Prostaglandin	PG2
Triunsaturated Prostaglandin	PG3
Cholesterol	CHOL

ENTER TEMPLATE NAME: _

Figure 1. Template selection in INPUT MODE.

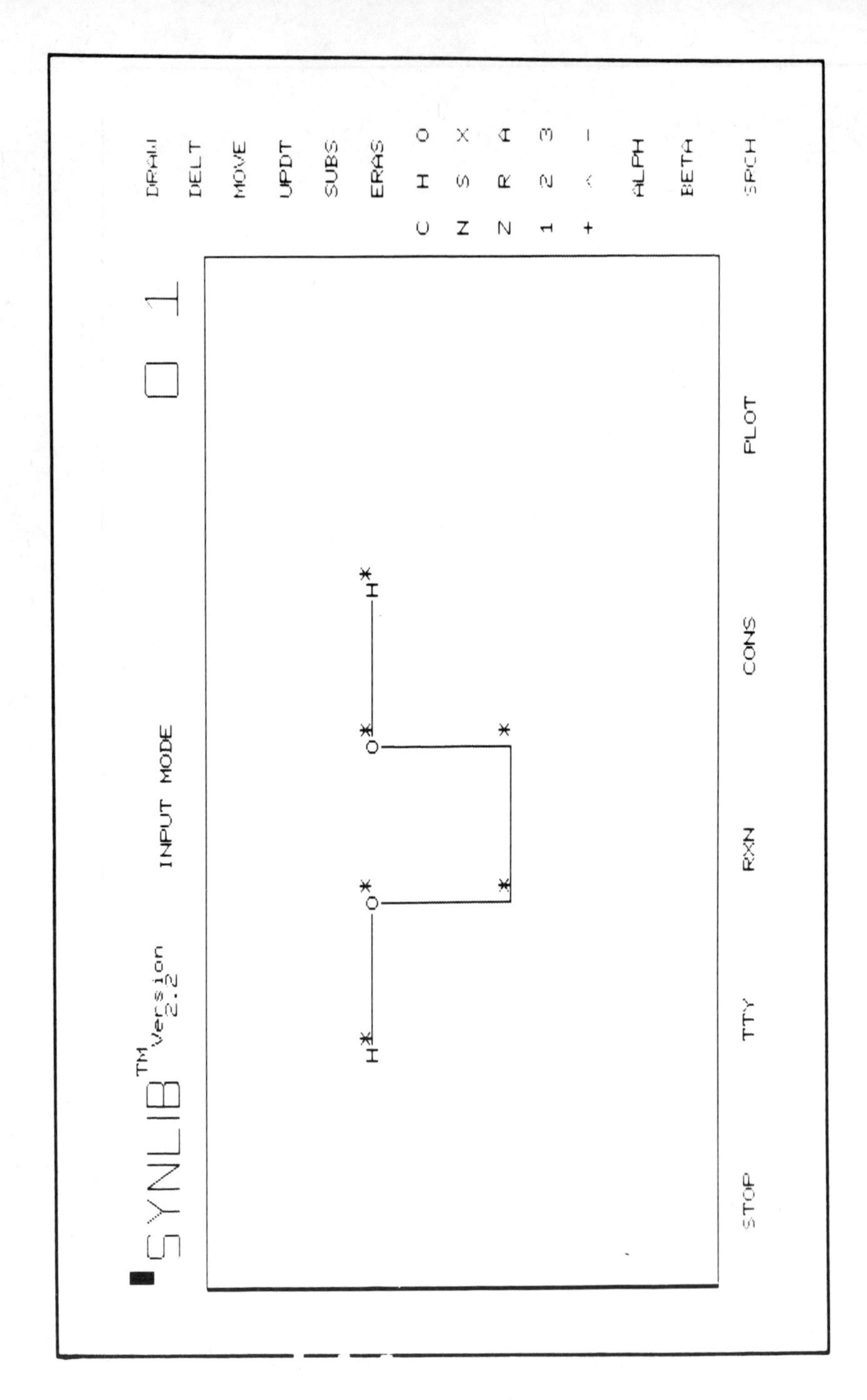

Figure 2. TARGET drawing in INPUT MODE.

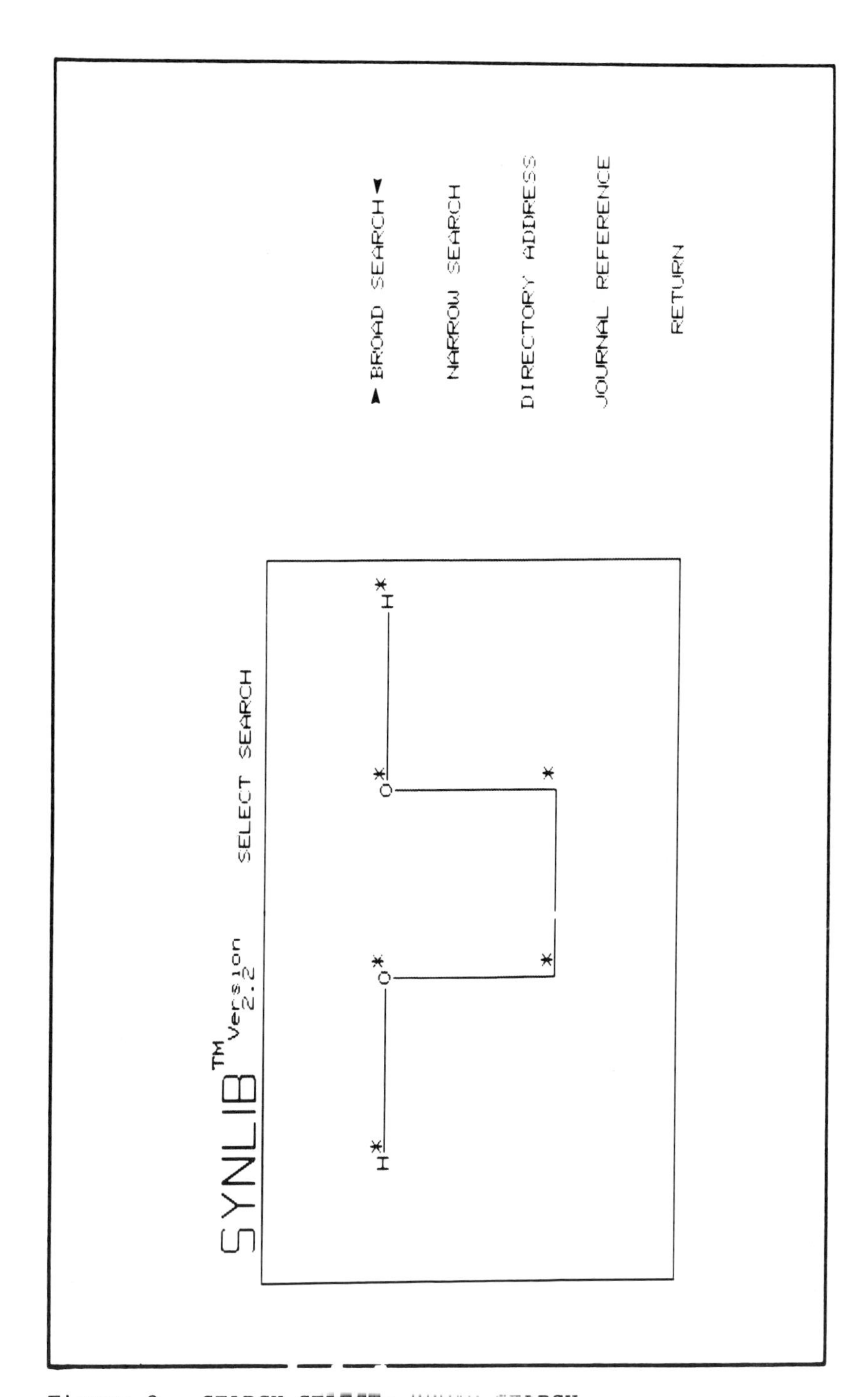

Figure 3. SEARCH SELECT: BROAD SEARCH.

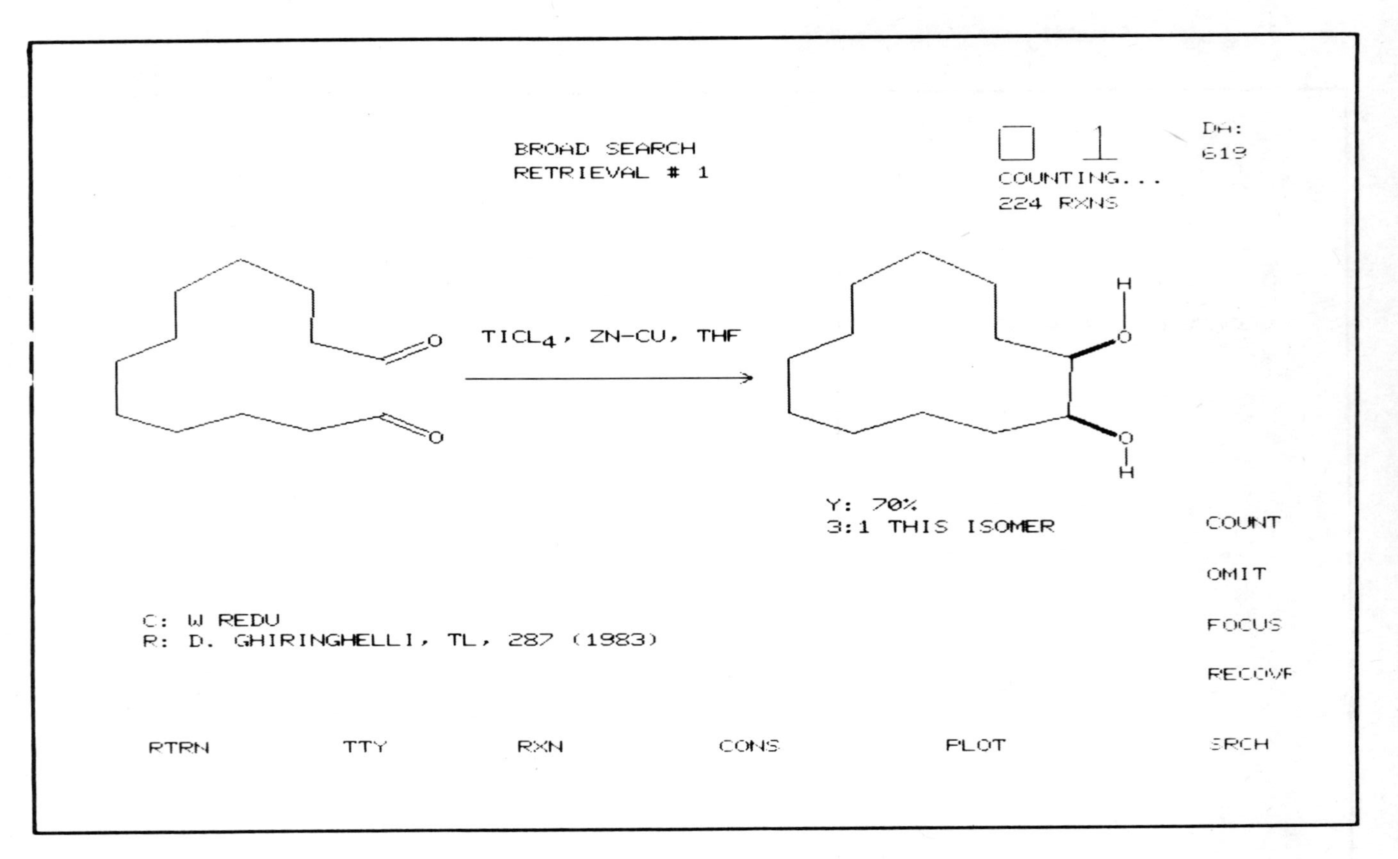

Figure 4. SEARCH MODE display of retrieved entry.

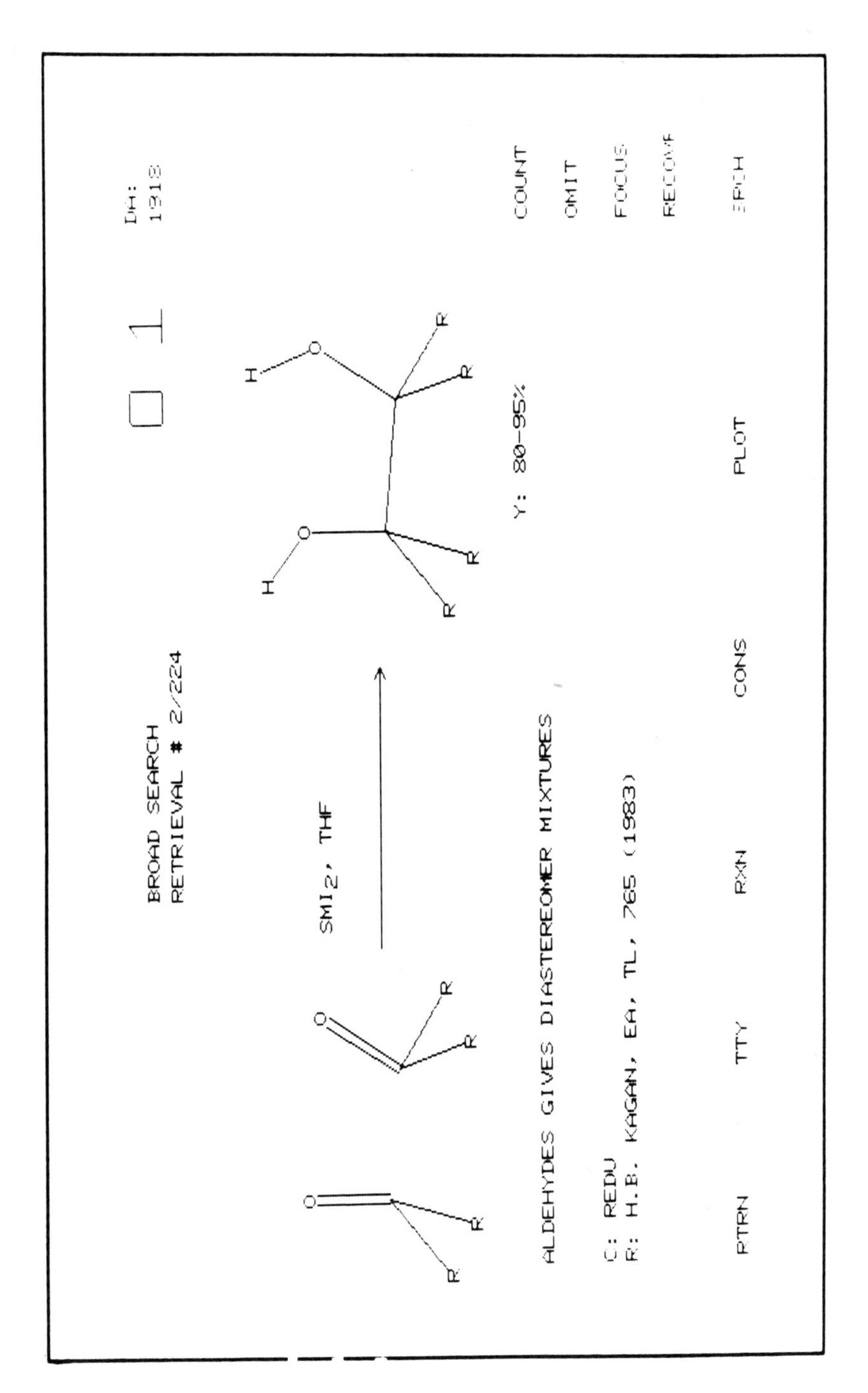

Figure 5. SEARCH MODE display of retrieved entry.

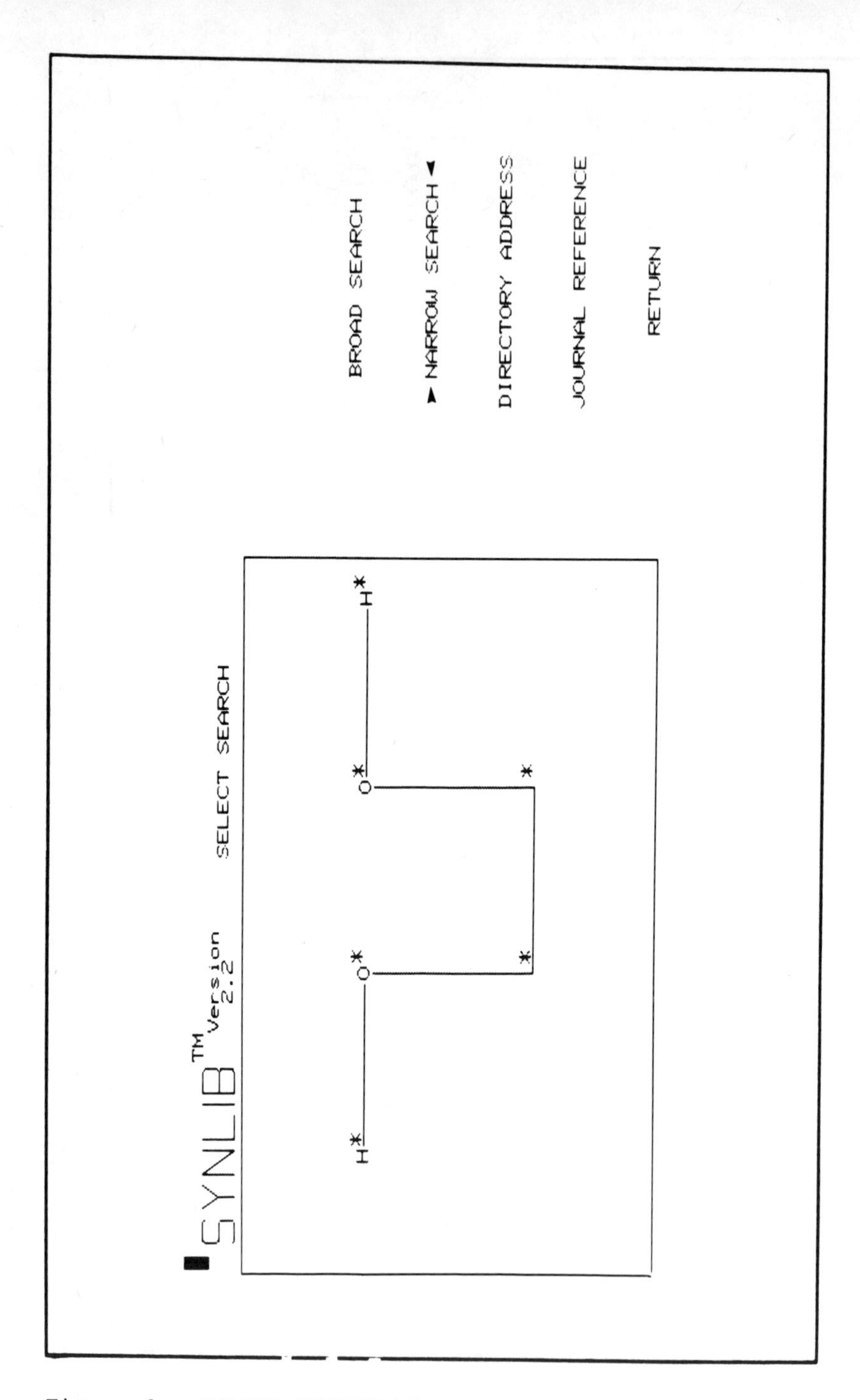

Figure 6. SEARCH SELECT MODE: NARROW SEARCH.

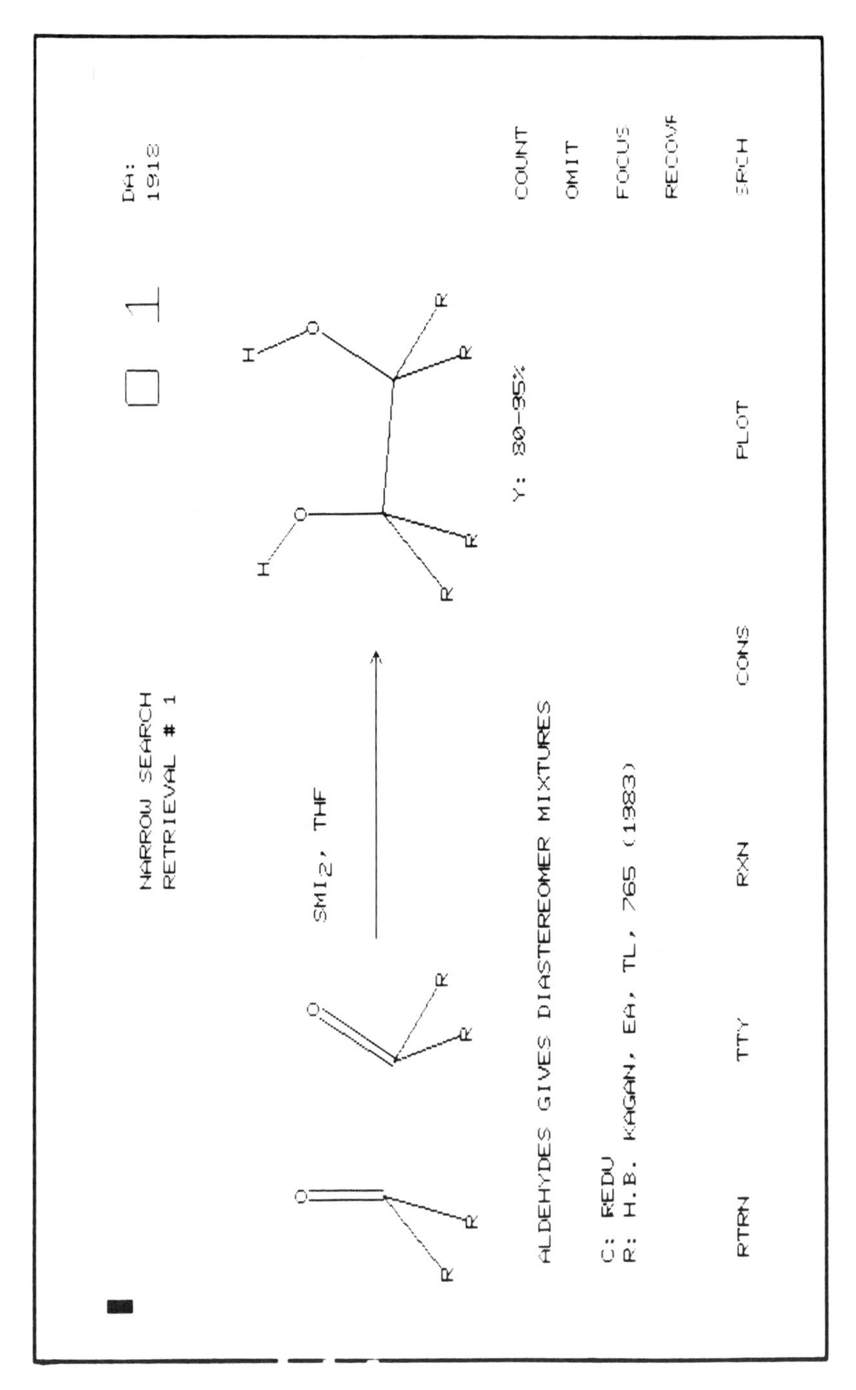

Figure 7. SEARCH MODE display of first matching entry.

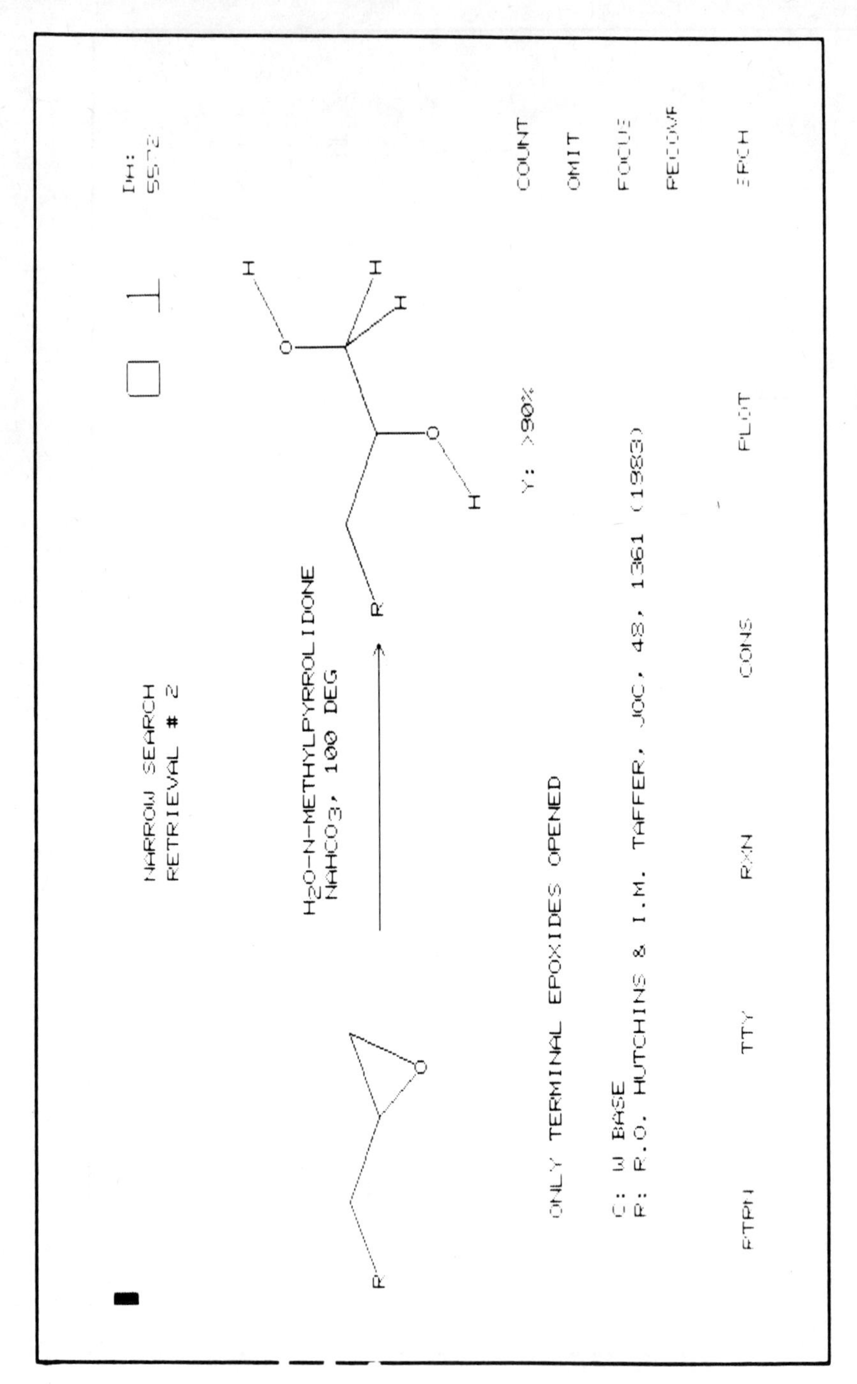

Figure 8. SEARCH MODE display of next matching entry.

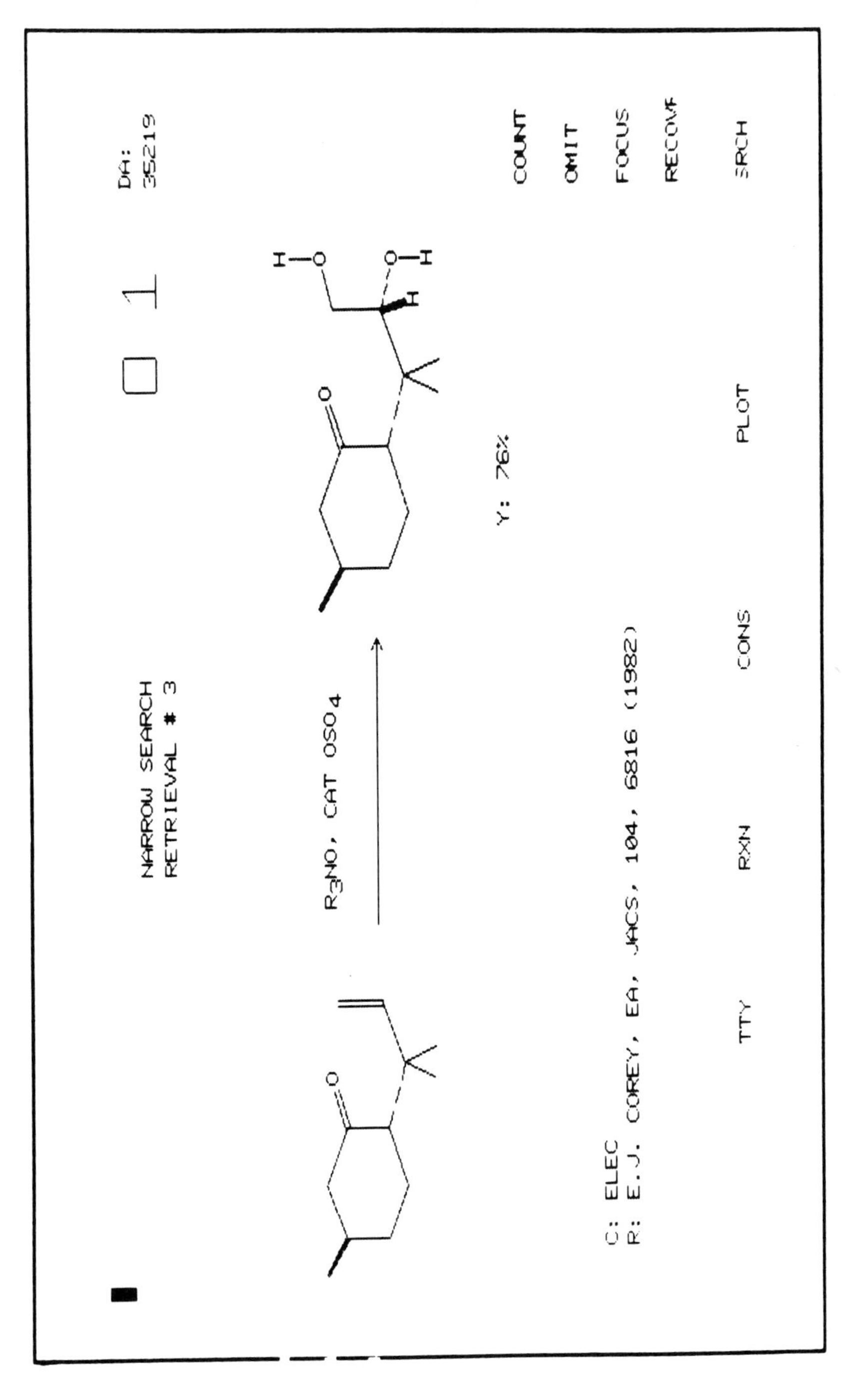

Figure 9. SEARCH MODE display of next matching entry.

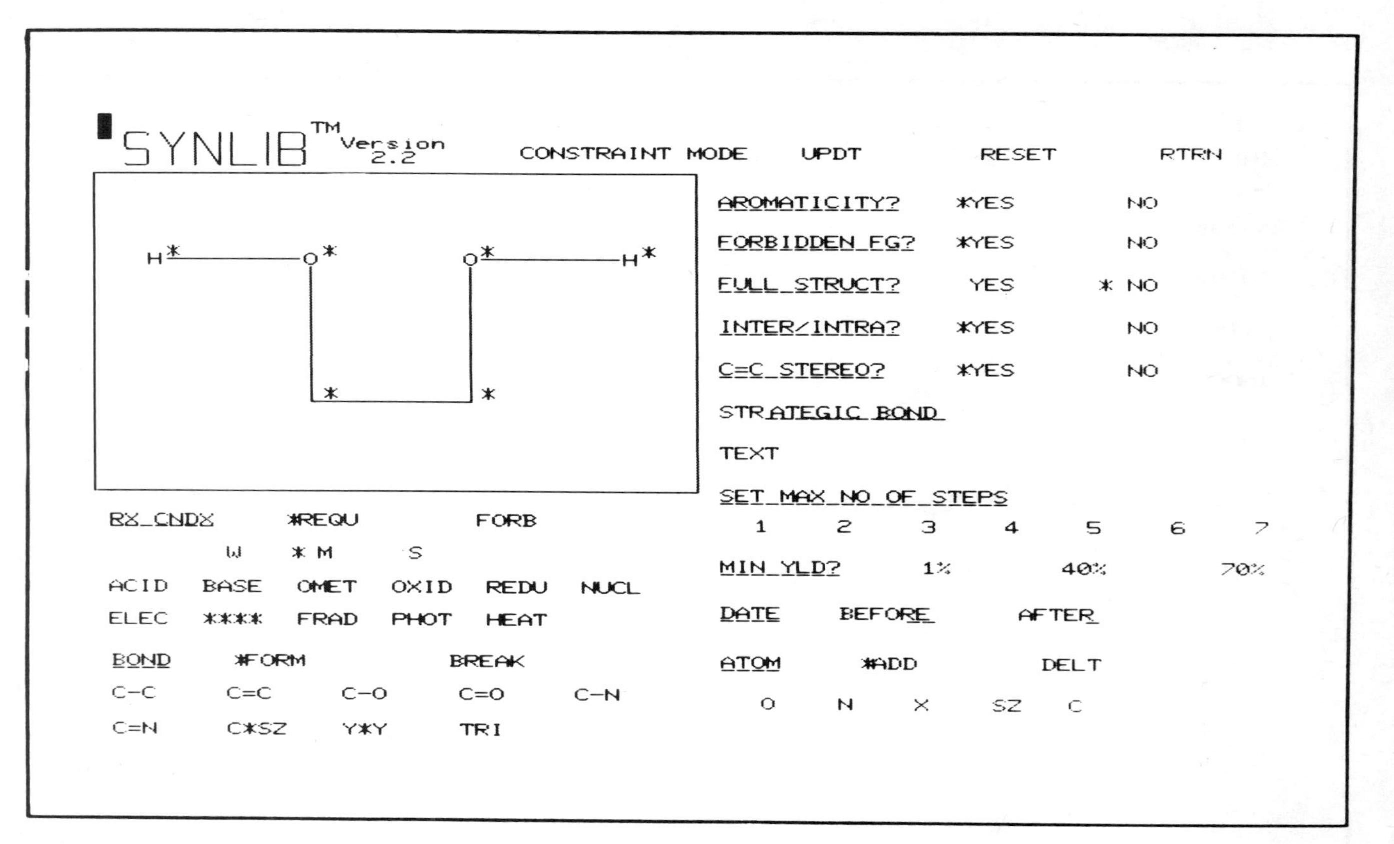

Figure 10. SYNLIB CONSTRAINT MODE screen.

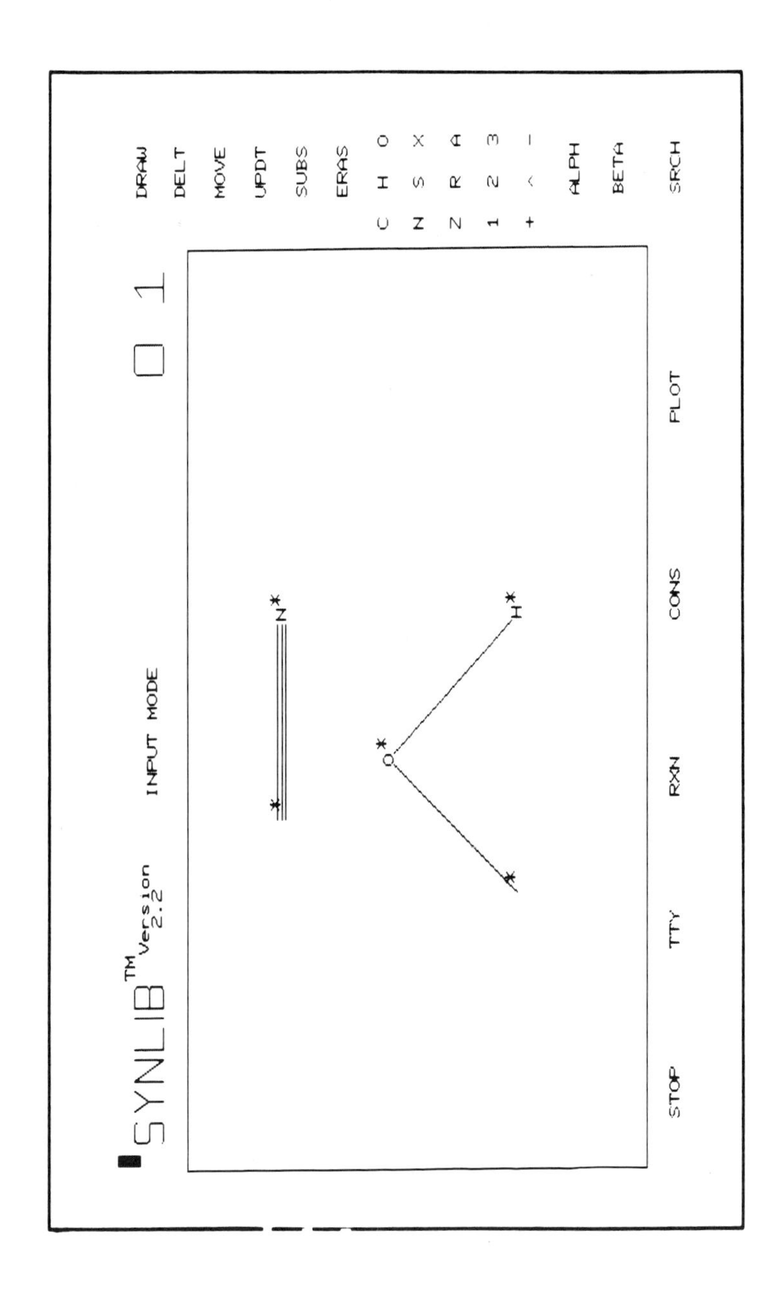

Figure 11. INPUT MODE is used to create the TARGET, two chemical structures that must each be independently located in the product structure. If the search is to be restricted, say, to primary alcohols the requisite hydrogen atoms should be explicitly drawn.

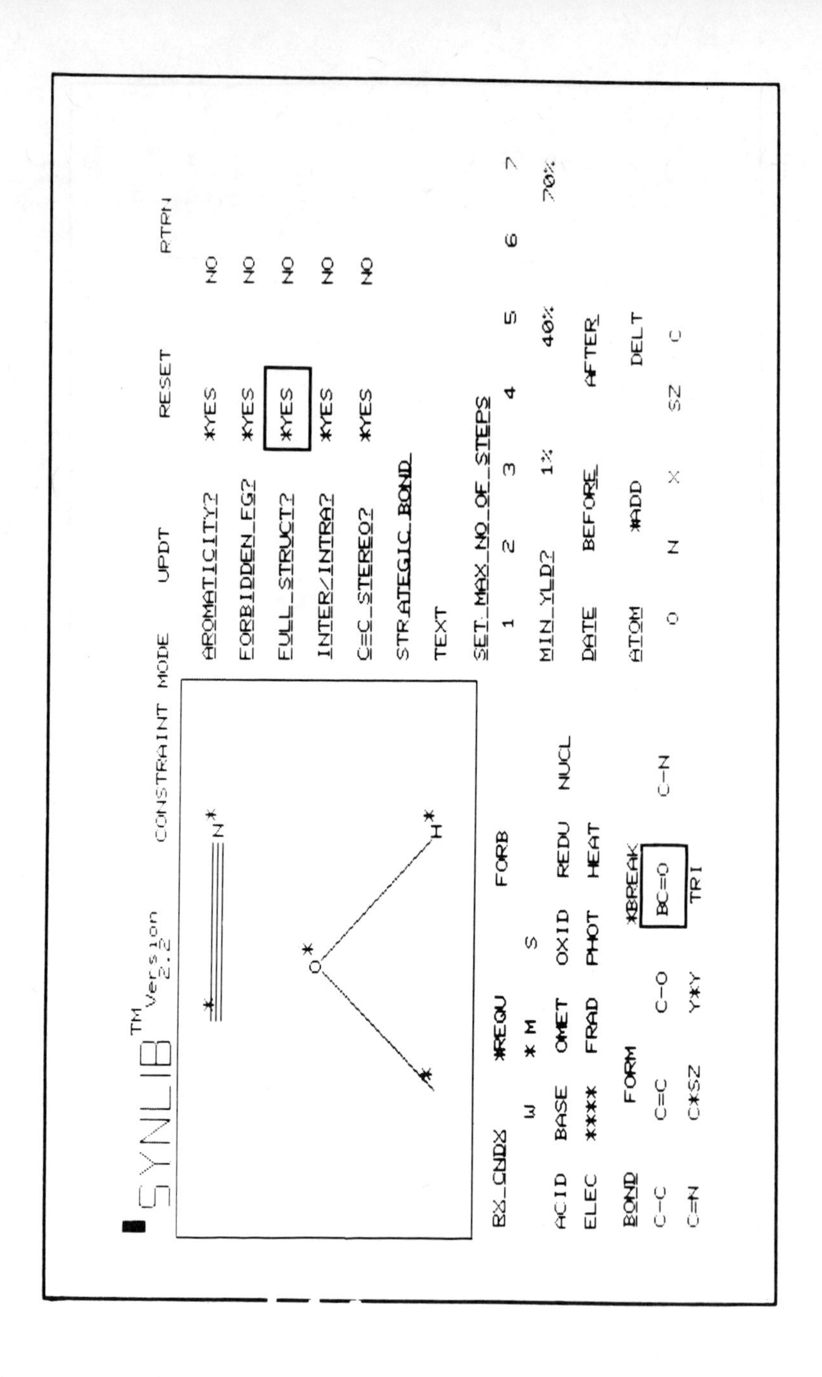

Figure 12. CONSTRAINT MODE is next used to select the FULL STRUCTURE search and to demand reactions that BREAK a C=O bond.

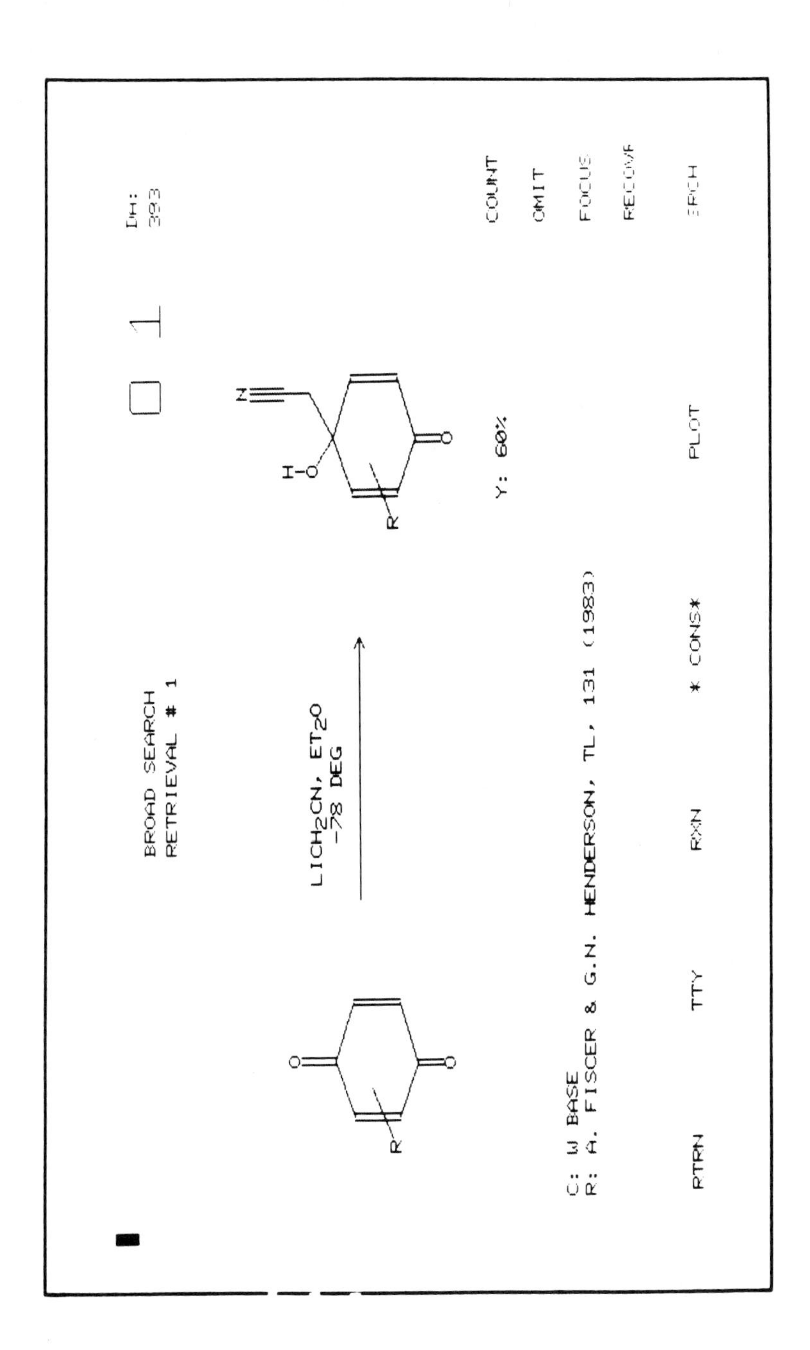

Figure 13. SEARCH MODE then executes the required search revealing the number of entries (using COUNT) awaiting display

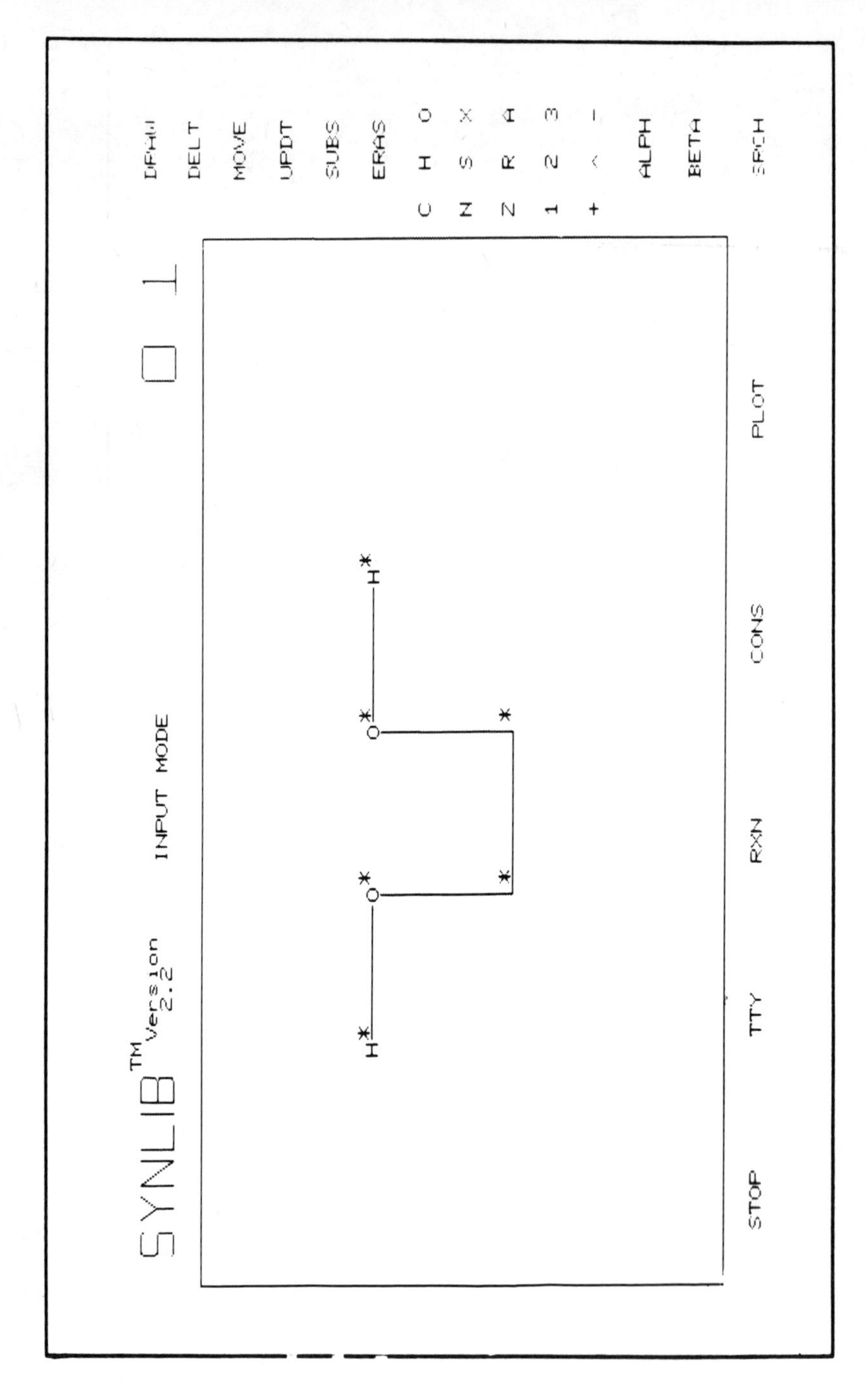

Figure 14. TARGET is drawn in INPUT MODE.

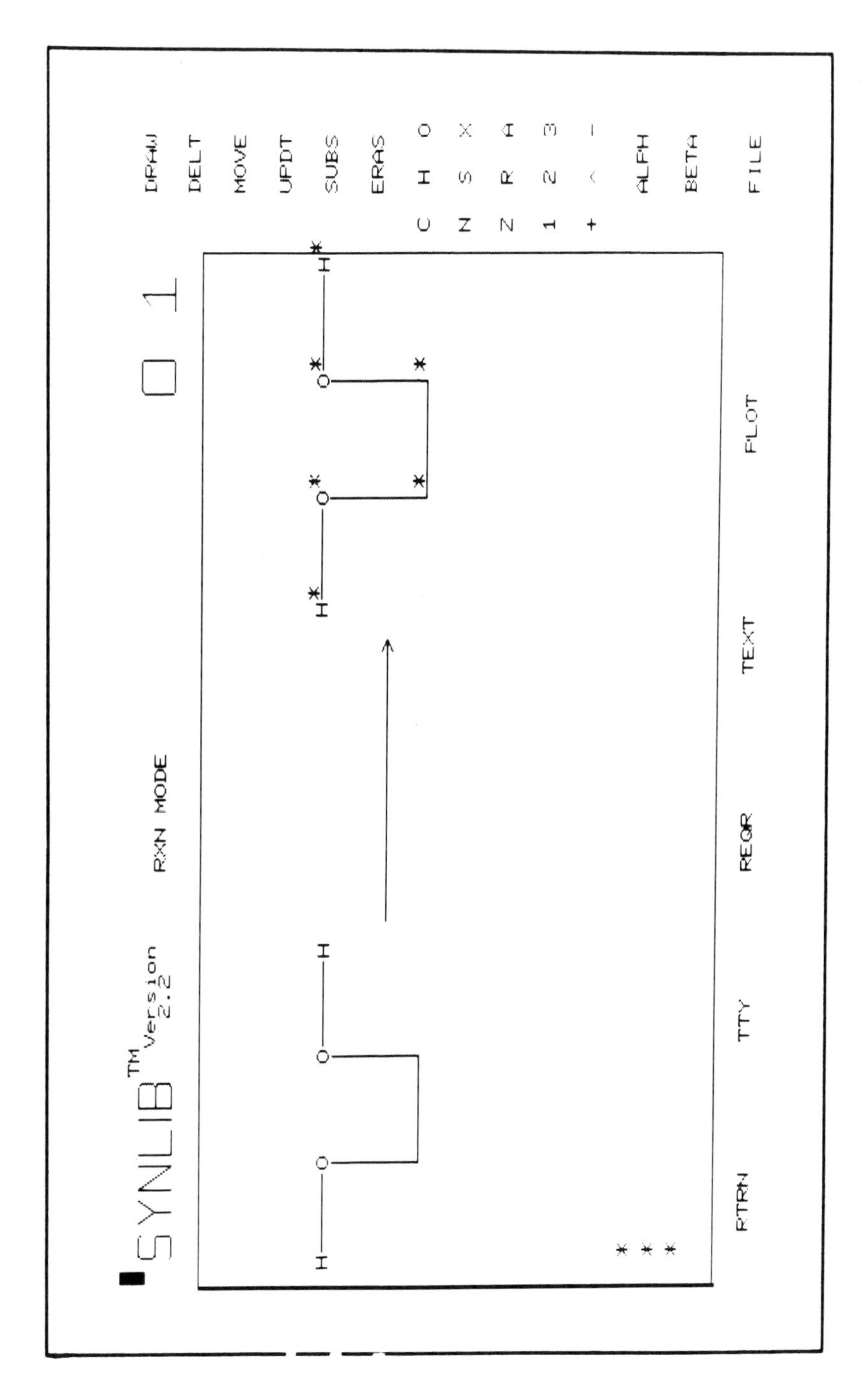

Figure 15. REACTION MODE is entered.

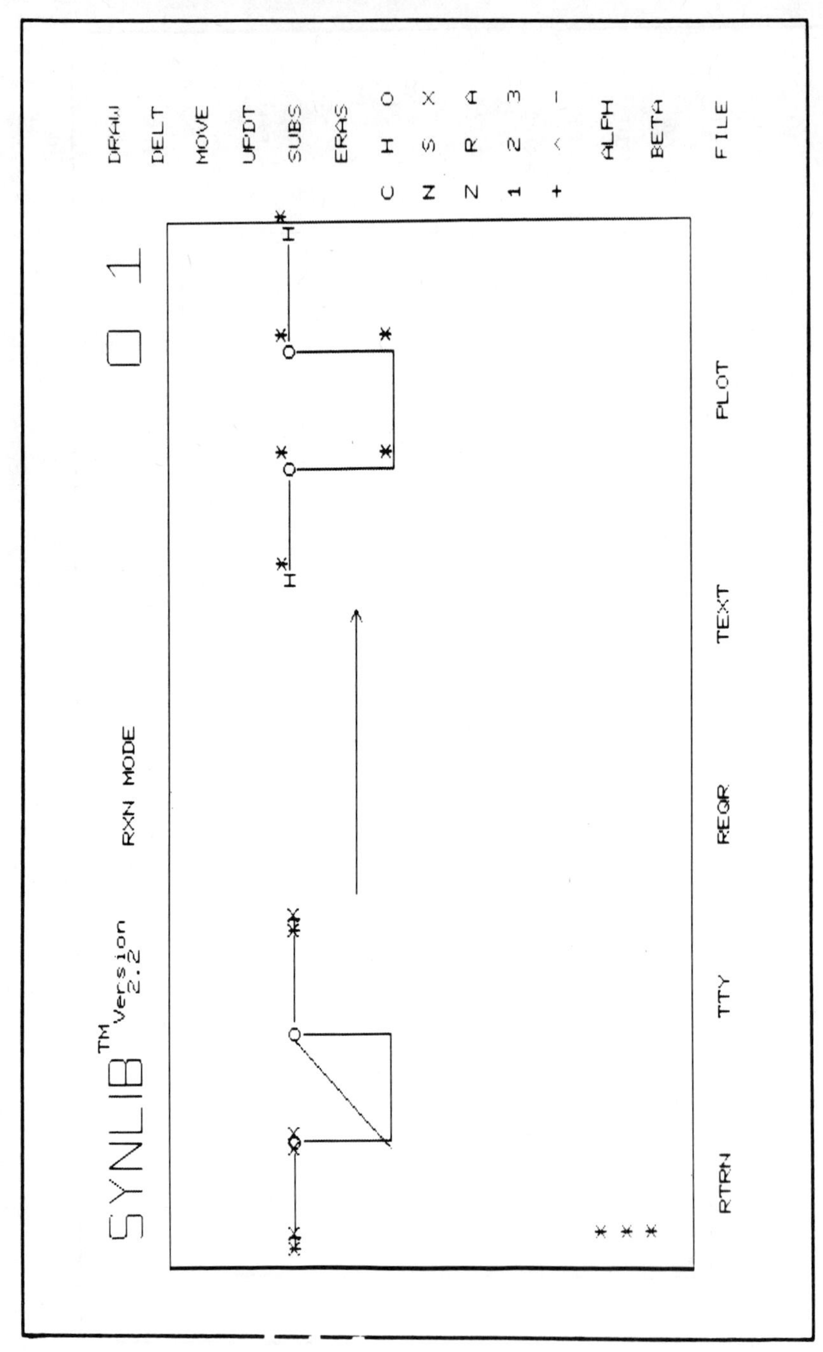

Figure 16. Starting material structure is edited as required.

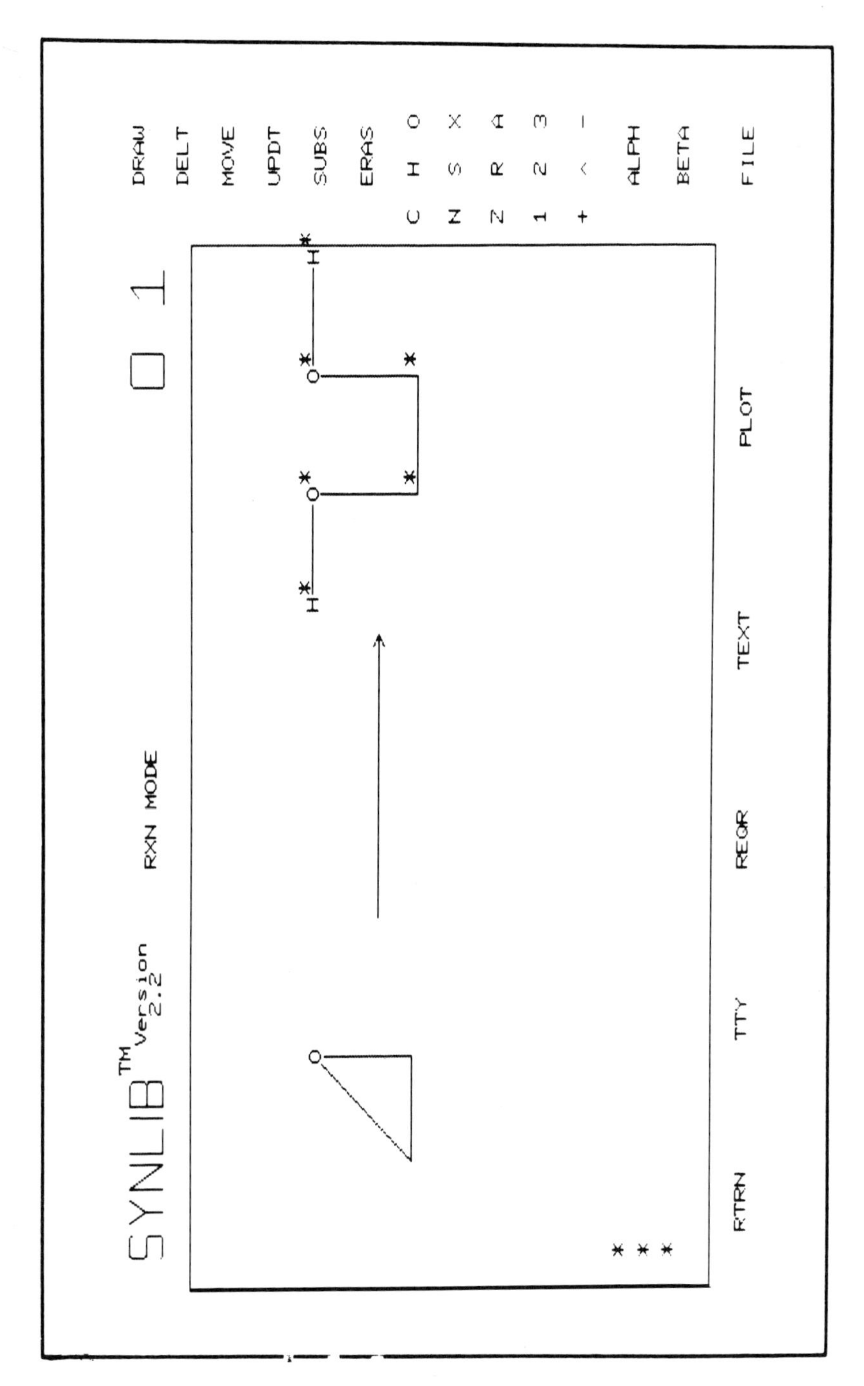

Figure 17. UPDATE redraws the reaction diagram.

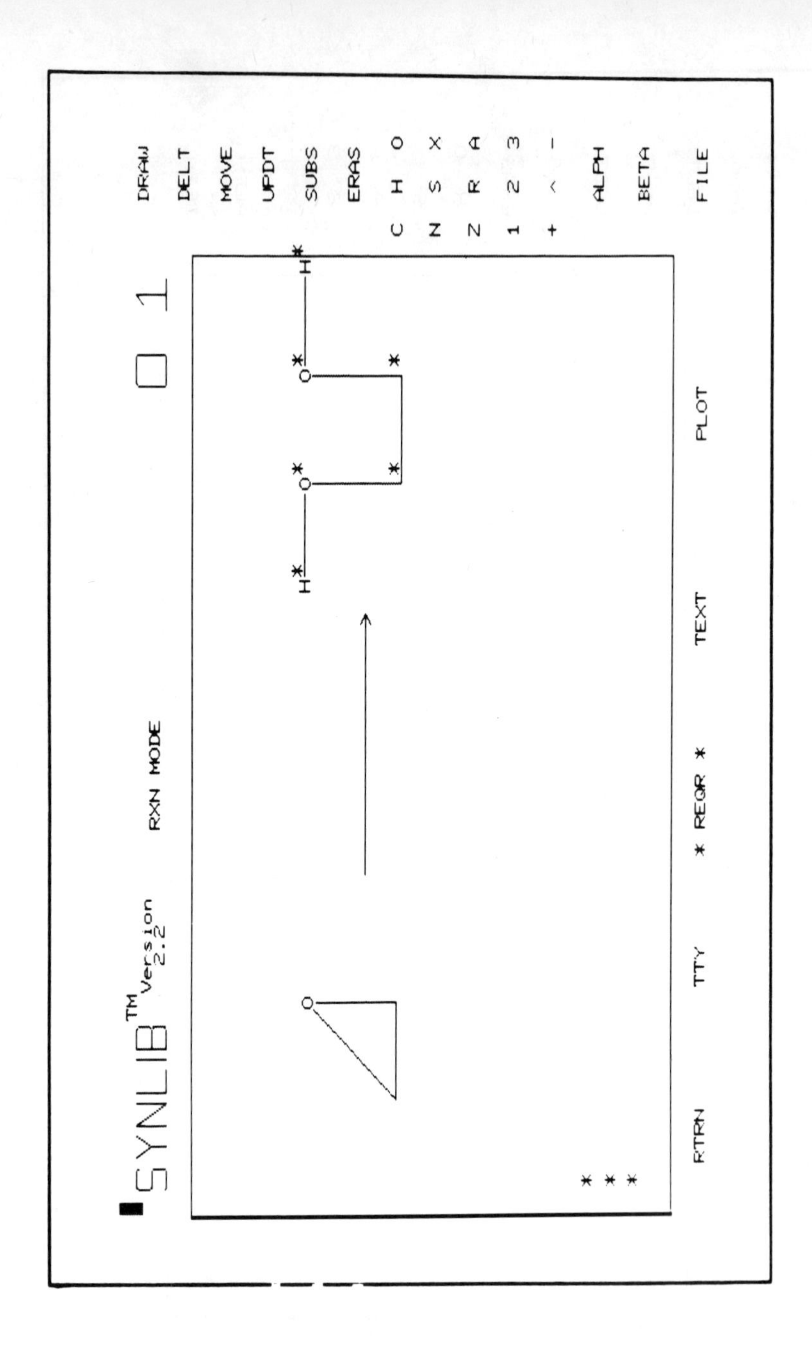

Figure 18. REQUIRE REACTION function is selected in REACTION MODE to constrain searches to reactions which add oxygen and yield products containing the glycol TARGET.

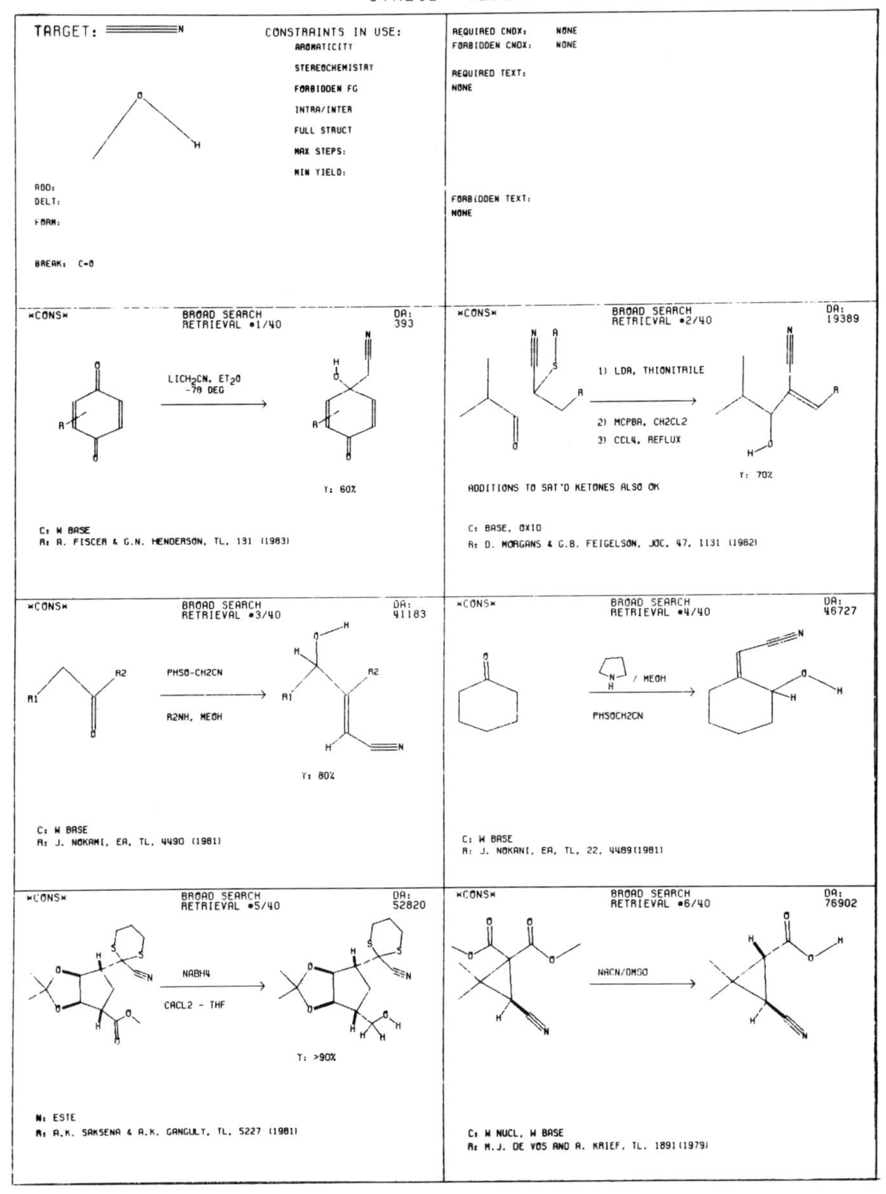

Figure 19. Typical search output produced by SYNLIB in a BROAD SEARCH.

11 Creation of a chemical reaction database from the primary literature

P. E. Blower and R. C. Dana
Chemical Abstracts Service

INTRODUCTION

Chemical Abstracts Service (CAS) recognises that the chemical community needs simpler and better methods than are currently available for searching the scientific and technical literature for chemical reactions. Development of the CAS ONLINE substance search software has put CAS in a position to develop the capabilities that will be needed for searching reaction information in ways that could not be conveniently done with a printed product. For the past several years, we have been examining users' needs for reaction information as well as the technical feasibility of providing a chemical reactions service. CAS has recently announced (1) its intention of offering an online Reaction Search Service and started building the database in October of 1984. Development of the systems for storing and searching reactions is now getting under way.

The service that we intend to offer will provide access to reactions cited in documents covered by Chemical Abstracts (CA) through CAS ONLINE, our online search service available through STN International. The initial system will allow users to search for chemical reactions from selected journals and patents by specifying structures or substructures and their roles in the reaction as reactant, product, reagent, catalyst or solvent. Users will also be able to specify the bonds in the product or reactant where the desired reaction occurs and only retrieve reactions occurring at that site. Providing such a service requires a database building component for constructing a computer-readable file of reaction records and a search component for retrieving selected records.

In this article, we will focus on the processes for creating a reaction database from information reported in the primary chemical literature. These processes involve a combination of editorial input and computer programs, integrated into the routine editorial indexing operations and CAS production processes. Figure 1 outlines the major processing steps for building a reactions database. Editorial staff are already adding reaction information as part of their normal indexing of chemical substances and editing it during routine editing of index entries. Subsequent computer processing will generate reaction records, identify reaction sites, and format reaction diagrams for editing and display. Finally, in order to ensure a high quality database, a second editing step will monitor and correct any errors in the reaction records. The verified reaction records will then be available for retrieval by users of our online search service. Before describing the details of this system and some of our initial experiences with it, we would like to consider the reaction information we have elected to include in the initial file.

FILE CONTENT

As we developed plans for an initial reaction file and search service, it was necessary to make several choices regarding file contents, both in terms of range of documents covered and of information selected from those documents. These choices required balancing at times conflicting priorities of user needs, production costs, and considerations of editorial quality control. Figure 2 summarises the choices we have for the initial reaction file. In the future, as we gain experience in the building and searching of this file, some expansions may be made; for example, one possibility might be to include papers from additional sections of CA.

The initial file will cover reactions of organic substances, including organometallics and biomolecules such as peptides, carbohydrates, steroids and prostaglandins taken from the organic sections (sections 21 - 34) of CA. At this time, approximately 100 core journals are being covered. In 1986, coverage will be extended to patents from eight U.S. and European patent offices (U.S., German Offenlegungsschriften, French, British, Swiss, Dutch, Belgian and European). We estimate that we will process approximately 16,000 documents in 1985 and increase to approximately 21,000 documents per year by 1986. By that time we will be encoding about 300,000 reactions per year.

Once a document is selected for reaction coding, there will be thorough coverage of the reactions described in it. There will also be thorough coverage of all substances participating in a reaction, including virtually all indexable catalysts, reagents, and solvents. These latter substances will be recorded both as specific substances and also given a generic classification. For example, lithium aluminium hydride would be recorded as the specific substance and also under the generic classification 'metal hydride'. In addition, the file will contain all multistep reactions that occur within a document. It will not include reaction yields, reaction temperatures, or substance quantities. The only types of reactions we are not recording from the documents selected for coverage are simple salt formations, the formation of characterising derivatives, and the formation of charge-transfer complexes, solvates, or hydrates. Some other specific reactions may be ignored if they are deemed of little use; e.g., the initial coupling of an amino acid to a solid support for a Merrifield synthesis.

REACTION DATA ENTRY

Present CAS ONLINE services - The Registry File and The CA File - involve providing online access to files that were already available. In the case of reaction searching, however, development of the service requires not only new search software but also development of a new and very large database. This database will require a continuing commitment from our editorial staff.

Document analysts in the Organic Chemistry Department at CAS are currently adding reaction data to the Chemical Substance Index entries. This data relates these substances to the particular reactions in which they participate and identifies the reaction role of each. This can best be explained in the larger context of document processing. In the course of indexing a document, the analyst numbers the substances that are to be entered in the Chemical Substance Index, and then creates an index entry for each substance. Consider the rearrangement reaction shown in Figure 3. In making the index entry for substance 1, for example, the analyst would cite its number and a descriptive phrase, which will appear with the index entry to indicate the nature of the information concerning that substance. In this case, the phrase (labelled Text Modifier) is 'cyanogen bromide mediated rearrangement of'.

The document analyst can enter the reaction information with the index entry for any substance participating in this reaction. This is simply a note identifying all of the reaction partners and indicating the role of each. We have chosen to show this as a note with substance 1 describing substance 2 as the reaction product, cyanogen bromide and potassium carbonate (substances 3 and 4) as reagents, and methanol (substance 5) as the solvent. If there were unusual reaction conditions, they would be described textually in an auxiliary reaction note.

Notice that the reaction information includes common reactants, reagents, catalysts and solvents. These substances are not normally indexed for CA. For example, methanol acting as a reaction solvent would not normally be indexed for CA. But we are including even such simple and common substances in the reaction file because we expect they will be useful for refining a reaction query. For example, such information would allow a user to focus on certain types of solvent systems or to screen out procedures involving hazardous or expensive substances. But inclusion of these substances in the reaction file does not mean a change in our editorial policy for CA. We have no plans for covering these substances in any other CAS product or service. That is the purpose of the Bypass Flag shown with the entries for substances 4 and 5 in Figure 3.

SOLUTIONS TO VARIOUS INPUT PROBLEMS

We have discovered a variety of problems as we started to build the database. In this section, we will describe some of these problems and the solutions we have used. In some cases, problems arose because of the way CAS handles and registers certain kinds of compounds. We will start with an illustration of this problem.

Coverage of 'double-entered' compounds. Certain kinds of compounds are 'double-entered' in CAS indexes. By that, we mean that they are given two different Registry Numbers although only a single compound is actually meant. This is generally done for nomenclature reasons. Two major classes of compounds which get such treatment are the ylides and the -atrane compounds. Figure 4 shows examples of compounds that are double-entered. We note here that the distinction, even in these broad classes, between single- and double-entered compounds is almost always one of author emphasis, in particular the representations chosen by the authors to describe such compounds in the original document.

Although the -atranes are relatively uncommon outside the organometallic section, the ylides are broadly used as reactants to convert carbonyl compounds to the corresponding olefins. They may or may not also be prepared as such, but more commonly they are generated in situ as the first step of the Wittig reaction. The most important characteristic of these compounds is that they have different connection tables and hence would be found by slightly different substructure searches. Therefore, we have decided to routinely enter both forms as though they were a mixture, and to associate the appropriate reaction role with each form. We realise that this may occasionally cause output in which there appears to be a mixture of reactants or products when in fact the original document only shows a single compound.

Compounds treated by CAS as General Subjects. Some compounds are treated by CAS as general subjects rather than substances, typically because some part of the molecule is not well-specified from the Registry System perspective. An example is the class of surfactants which incorporate alkyl moieties derived from a natural source, e.g., coconut oil. Although these alkyl groups may be reproducible in the sense that they can be used to make the same compound repeatedly, they are ill-defined for the Registry System since they typically include a variety of chain-lengths and the possibility of unspecified unsaturation. In a given series of trade-named surfactants, some will be registrable compounds and others will receive only general subject coverage. The analyst will typically not know at the time of analysis which substances get which treatment.

This is a problem solved in the course of our initial online editing. When the Document Processing System discovers that a nominal compound entry should be treated as a general subject entry, the online editor deletes the compound but adds a Reaction Note citing it as a reaction participant. The note will contain the candidate name for this 'compound'. Although this note was originally designed as a way to reflect unindexable reaction conditions, it is also the best practical way to handle these nominal compounds.

Failed reactions and trivial compounds. We will consider failed reactions and trivial compounds together, although they have little in common conceptually, because in both cases the reaction indexing involves species which we would not otherwise have indexed. Yet because there is a significant chance that such reactions will involve new compounds, we are not able to use the Bypass flag noted above for common compounds.

Authors will sometimes find that an expected reaction simply does not work. This may occur at the end of a long and laborious multi-step synthesis of a naturally-occurring compound. For a variety of reasons, primarily the author's emphasis and interest in the compound which could not be prepared, CAS has decided to include such failed reactions in this new database. We believe that there is value in such failed reactions, but at the same time we want to make it clear that this reaction did not work. Therefore, we have associated with such reactions both Text Modifications and Reaction Notes which state that the reaction failed.

We would like to emphasise here that we will not index as a failed reaction all reactions in which no product was found. To be included as a failed reaction, there must be some author interest in and emphasis on the failure of the reaction. And the author must specify the expected product. We have covered such reactions in the past by use of a Text Modifications which indicated attempted reactions with the index entry of the reactant(s). But we have not previously indexed the products in such reactions.

Our treatment of trivial compounds also represents a change in practice, although not in policy. For many years, analysts at CAS have ignored compounds which were made in very low yield, unless those compounds are the focal point of the paper. We have generally not indexed compounds present in less than five per cent yields. However, for both aesthetic and systematic reasons, CAS decided to include these relatively trivial compounds in the reaction database. We are also, to some extent, indexing the by-products from reactions. This does not mean that we will start to index the carbon dioxide, water or inorganic salts given off in elimination reactions. But we will include some significant organic moieties that we previously ignored because the author identified them as by-products.

'Recreated' reactions and reactants. One of the best ways to shorten a preparative patent or journal article is to record syntheses by referring to previous examples. Thus, an author will show a 'typical' synthesis in some detail and then give a table of compounds which were prepared 'similarly'. We have some difficulty discovering the extent of the 'similarily'. CAS has, for some time, 'recreated' the reactants for those reported products when there was no ambiguity associated with the reaction. By that, we mean that there are no unspecified leaving groups or anything else which makes one or more reactants open to question. In practice, we have also not recreated multi-step reactions back to the initial starting materials. For the reaction service, we are attempting to include recreated single-step reactions which are unambiguous; we are less inclined to include multi-step reactions because of the large number of compounds involved.

Sometimes the author alters this procedure slightly by showing the 'major' reactant together with the product, but the minor reactant is not specified. Generally, this is so because of an unspecified leaving group; most common examples are acyl and alkyl halides. Although we can easily tell from the product the nature of the acyl or alkyl group, the halide cannot be determined. Yet these may be interesting reactions, so we have decided to create default halides for these cases. In general, we will assume that acyl halides are the chlorides and that alkyl halides are the bromides. But when we make such assumptions, we will always write a Reaction Note indicating that the halide was actually not specified.

In both cases described above, we recognise that this information is not explicit in the actual document. And we have been been reluctant in some cases to make a conjecture about what species were actually present when preparations were done 'similarly'. Therefore, we have refrained in some instances from specifying solvents or other reaction participants which are less critical to the reaction. We have concentrated instead on linking important reactants and products. Because our document analysts are well-versed in their special areas, we feel comfortable in our assumption of unspecified reactants and recreated reactions for inclusion in this database.

We would just like to mention in passing a synthetic method that is quite common but is causing us some problems;

this is the Merrifield synthesis of peptides. For those who may not be familiar with this method, it is a convenient, clean way to make large peptides; in fact, it is clean enough that it has been automated. Generally, the first amino acid in the peptide is N-blocked and attached through its carboxylic acid group to some solid support such as a suitably functionalised polystyrene. Reagents are added to deblock the amino acid and then an appropriately-protected second amino acid is introduced, resulting in coupling with the amino acid already there. This sequence of amino acid coupling, deblocking and further coupling can be repeated a large number of times.

We have compromised in our coverage of these Merrifield syntheses. Since it is usually the sequence of the final peptide that is of interest, we show the addition of all reacting amino acids explicitly. But we only describe the deblocking and coupling conditions for the first step. We are now investigating the limits on this coverage, particularly in terms of the size of the resulting peptide.

IDENTIFICATION OF REACTION SITES

Reaction queries often contain specifications that certain bonds be formed or broken because the chemist is interested in a very definite type of structural change. So the search system must be able to determine which reactant and product bonds were involved in a reaction. We have developed a computer program to identify reacting bonds by examining the structure records for the reactants and products. This program simplifies reaction input considerably. Recall that when a document analyst enters a reaction he refers to substances by numbers. At that time, there is no structure diagram or connection table which he could use to enter reaction site information. There does not appear to be any convenient and reliable way for the document analyst to enter this information without reprocessing the document at a later stage.

The reaction site identification program uses a matching procedure to identify the structural differences between the reactants and products. It is based on work by Vleduts (2-3) and Willett and Lynch (4-5). The basic heuristic used for identification of reaction sites is the assumption (3) that the largest substructure common to the reactants and products is the unchanged portion of these substances. Thus, the reacting bonds are those bonds adjacent to this common substructure but not contained in it. In other words, a bond is in the reaction site if it is not in the common substructure, but one of its atoms is in the common substructure.

The program for identifying the common substructure and the reacting bonds operates in two phases. Phase I searches for a minimal set of bonds that can be removed from the reactant and product structures so as to afford a substructural match at the graph/node level. By that, we mean that the match includes topology and atom type but not bond type. This bond set corresponds to the bonds that are completely formed or completely broken in the reaction. Phase II identifies the reacting bonds that only changed in bond value (e.g., from double to single) and bonds to hydrogen atoms. Phase II is a straightforward procedure for refining the results obtained in the first phase. The bulk of the processing takes place in Phase I which contains subprocesses for matching reactants and products and for identifying candidate reaction sites.

The first step in Phase I is an attempted match of the reactants and products. The matching procedure employed is an adaptation of Algorithm I of Corneil and Gottlieb (6) and uses a set-reduction technique. Initially, reactant and product atoms are assigned to sets based on atom type, and the sets are assigned sequential class values. Thus, at the outset, all carbon atoms have one class value, all oxygen atoms have a second class value, etc. Thereafter, each iteration of the matching algorithm refines or partitions the sets based on the class values of the immediate neighbours. This partitioning can be explained as follows: Consider two atoms X and Y from the same class after the *i*th iteration. Let CV(X) and CV(Y) denote the sorted lists of class values of the immediate neighbours of X and Y. Then X and Y will be in the same class after the (*i+1*)-th iteration if, and only if, CV(X) = CV(Y). After each partitioning, the procedure reassigns class values to the new atom sets. Classes containing both reactant and product atoms are called *active*; all other classes are inactive.

This refinement procedure is illustrated in Figure 5 for the bromination of toluene. At each iteration, the atoms in active classes are labelled with their class values, and the atoms in inactive classes are labelled with an asterisk. The reaction diagram at the top shows the initial situation. All carbons are in class 1, and the bromine is in class 2. The bromine is labelled with an asterisk, rather than a 2, to indicate that there is no matching atom in the reactant. After the first iteration, all carbons adjacent to exactly three other carbons are in class 1, those adjacent to exactly two other carbons are in class 2, etc. The product carbon bearing the bromine is now labelled with an asterisk because there is no reactant atom in a similar environment.

The matching procedure is terminated by one of two conditions: (1) At some iteration level *n*, one of the reacting substances has no atoms in an active class. (2) The number of classes at iteration levels *n* and *n+1* is the same; this indicates that no further refinement of classes can be achieved. In the example of Figure 5, the procedure terminates after the fifth iteration because there are no active classes.

The asterisk atom labels are intended to draw attention to the way atom classes are broken up as the matching proceeds through successive iterations. The bromination of toluene evidently involves the replacement of a C-H bond by a C-Br bond, so the structures differ at that carbon. Looking at Figure 5, we can see that at each iteration of the matching procedure, this difference in the two structures is propagated one bond at a time. The second stage of Phase 1 uses this information which is gathered during the matching process. For each atom, the matching procedure records the highest iteration level at which the atom was a member of an active class and its class value at that point.

The second stage of Phase I is the location of corresponding reaction site atoms of the reactants and products. Reactant and product atoms in the same class at the highest iteration level should be corresponding atoms in the common, unchanged substructure (4); for example, the methyl groups in the bromination of toluene. Furthermore, if *n* iterations were required before a complete mismatch was encountered, these atoms should be *n* bonds from a reaction site. So the program generates pairs of paths of length *n* starting from corresponding reactant and product atoms. At any atom *A* in the path, the path will proceed from *A* to the neighbour of *A* with the lowest level. If there is more than one choice at atom *A*, then all such paths are generated. Finally, the path-pairs are ranked, with higher priority given to paths which appear to pass through corresponding atoms and lead to reaction site atoms. For example, Figure 6 shows the highest ranking paths for the cyanogen bromide rearrangement considered earlier. These paths, indicated by heavy shading, proceed from the methoxy groups to the atoms labelled *R* and *P* in the reactant and product, respectively.

The third stage of Phase I is the selection of candidate reacting bondsets. The terminal reactant atom *R* and product atom *P* on any path-pair generated by the procedure just described are assumed to be corresponding reaction site atoms. That is: local differences in the environments of *R* and *P* are due to bond changes resulting from the reaction. This procedure generates all possible ways of removing bonds to atoms *R* or *P* so that their local environments will become the same. In Figure 6, for example, both terminal atoms have two carbons and an oxygen neighbour. But the product atom has a second oxygen neighbour, while the reactant atom has a nitrogen neighbour. Since either C-O bond might be the reacting bond, there will be two bondsets generated each containing one of these alternative C-O bonds. Both bondsets will contain the reactant C-N bond.

The search for a minimal set of reacting bonds is performed by a backtracking, tree search procedure. Each node in this search tree corresponds to an alternative set of reacting bonds. And the three operations just described - structure matching, location of reaction site atoms, and generation of alternative bondsets - are common actions taken at each node in the search tree. These operations result in a set of alternative bondsets, and each of these bondsets corresponds to a new branch in the tree. For any bondset, the bonds listed are removed from the reactant and/or product structures, and the procedure is repeated. For example, Figure 7 shows the search tree and the bondsets generated for the cyanogen bromide rearrangement. In this tree, left branches represent successor bondsets or 'sons' and right branches represent alternative bondsets or 'brothers'.

This recursion is terminated in one of two ways: (1) The matching procedure finds a substructural match. In this case, the current set of reacting bonds is taken as the minimal set, and the program backs up to search for a smaller set. (2) The program fails to find an acceptable match. The program is run with an adjustable parameter that indicates the maximum number of bonds that can be removed in searching for a match. Currently, the parameter is set at ten. This seems to be about right since most reactions only involve a small number of bond changes. This search procedure constitutes Phase I of the program and results in a set of reacting bonds. The reacting bonds at the termination of Phase I are marked by a slash in Figure 8.

Phase II identifies the reacting bonds that only changed in bond value (e.g., from double to single) and bonds to hydrogen that are formed or broken. A by-product of the matching process is a reactant to product atom map. And Phase II simply compares the bond values of corresponding bonds and the hydrogen counts of corresponding atoms and records the differences. The one additional reacting bond identified in Phase II is a methyl C-H bond. This methyl group is labelled with an asterisk in Figure 8.

The current version of the program has been tested on several thousand reactions created during our input experiments. An error analysis suggests that the program will be able to successfully process about 92 per cent of the reactions. This program also checks the results for potential errors and, if appropriate, adds codes to assist in manual resolution of errors. Records containing potential errors will be corrected at a later stage of the processing.

QUALITY CONTROL AND ITS IMPACTS

Analyses performed during experimental input of reactions indicate that we should expect errors in reaction records to arise from several sources. Some will be input errors, such as invalid or missing substances. Other errors will be caused by the reaction site identification program, (although that program will also flag records which appear to contain invalid or suspicious data). In addition, errors will also be uncovered by internal and external users of the file.

Although many of these errors will be minor and will not necessarily affect search results, we want to eliminate as many errors as possible. So, as a final quality control step, we plan to develop an Online Reaction Edit program for monitoring and correcting errors. This program will allow an editor to review reactions suspected of containing errors and make any necessary corrections to reaction sites, substance identity or reaction role assignments. When reaction records have passed this quality control step, they will be added to the main reaction file and will become available for retrieval by users of the reaction search service.

Since 1972 when CAS began to integrate the abstracting and indexing functions so that there would be a single intellectual analysis of the document, there has also been an attempt to make the initial input as good as possible. Although highly qualified specialists will review the analysis twice before it is ready for publication, our emphasis has been on getting it right the first time. We have also installed a variety of specialised edits to detect errors or output diagnostics to focus the attention of subsequent editors. The Reaction Search Service now has its own set of specialised edits. One set examines the input data in preparation for online editing by CAS chemists. A second set of edits is incorporated in the reaction site program, but these edits cannot be performed until substance registration is complete.

The pre-edit processes performed prior to normal online editing by chemists are aimed at perfecting the mechanism by which the reaction participants are linked together in the database. Thus, they indicate that participants specified for a reaction have no corresponding index entry, highlight the absence of solvent, or indicate that an appropriate reaction role could not be assigned. All of these edits are designed to smooth and strengthen the normal editing involved in preparation of the CAS database.

During the process of locating reaction sites, a second series of edits is performed. Since this process involves a detailed comparison of the reactants and products, it is likely to uncover errors that would be difficult to detect otherwise. However, these programs run well after the normal editing cycle is complete. Thus, it has been necessary to build a separate editing process involving both online resources and hard-copy alerting. The actual problem detected by these programs is output in graphic form on hard-copy. Since the solution to the problem may require that several of the original indexing workunits be changed, the entire indexing for the document must be available online.

The Reaction Service is not tied explicitly to publication of the 'normal' CAS database. We estimate that reactions from a given document that is abstracted and indexed in the normal database may not appear in the Reactions Database for approximately a week. This delay arises primarily because of our need for extensive processing after connection tables for all reaction participants are available. The errors found in this stage are expected to be processed rapidly, but some errors may require considerable time to resolve. For example, if the system determines that a major reactant has been missed, the editor must (1) determine that the compound is indeed missing from the indexing, (2) add the compound as a new index entry if necessary, and (3) correct any reaction information involving the missing reactant. For compound-intensive documents, this may be a time-consuming task.

Upgrades to registered compounds which are involved in the reaction database will also trigger the system to check for impacts on the reactions. Thus, if the stereo text descriptor of a naturally-occurring compound is updated, the system will check to make sure that this does not change the work reported in the reactions database. Since the timing of such changes cannot be predicted, the system must be able to respond at any time, perhaps even years after the reaction was initially placed in the database.

CONCLUSION

The important processes and programs needed for production of a reactions file have been tested extensively by CAS. So although the creation of a reaction file will require a significant effort from our editorial staff and some program development remains to be done, we do not anticipate any major obstacles in installing these procedures in our current manufacturing system.

In conclusion, we believe that with this file, CAS can offer a reaction service with a combination of attractive features. The file will provide broad coverage of the reaction literature. The initial file will cover organic reactions, and we intend to extend the service to other CA sections after we have a clearer understanding of users' needs for reaction information.

REFERENCES AND NOTES

1. P.E. Blower, Jr., 'Design considerations for a chemical reaction search service,' 188th American Chemical Society National Meeting, Philadelphia, PA, August 30, 1984.

2. G.E. Vleduts, 'Concerning one system of classification and codification of organic reactions', Inform. Stor. Retr., 1, 117, 1963.

3. G.E. Vleduts, 'Development of a combined WLN/CTR multi-level approach to the algorithmic analysis of chemical reactions in view of their automatic indexing.' British Library, R&D Dept. Report 5399, 1977.

4. M.F. Lynch and P. Willett, 'The automatic detection of chemical reaction sites', J. Chem. Inf. and Comp. Sci., 18, 154, 1978.

5. J.J. McGregor and P. Willett, 'Use of a maximal common subgraph algorithm in the automatic identification of the ostensible bond changes occurring in chemical reactions', J. Chem. Inf. and Comp. Sci., 21, 137, 1981.

6. D.G. Corneil and C.C. Gottlieb, 'An efficient algorithm for graph isomorphism', J. Assoc. Comput. Mach., 17, 51, 1970.

Editorial Input

- Reaction Notes Input with Index Entries.
- Online Editing of Reaction Notes.

Computer Preparation of Reaction Records

- Interpret information supplied during document analysis.
- Indentify reaction sites in reactants and products.
- Format reaction diagram for editing and display.

Quality Control.

Figure 1. Reaction File Building Processes.

Number of Documents per year	21,500
Number of Reactions per year	300,000
Type of Documents	Journals - 100 core Patents - US, Germany, France, Great Britain, Switzerland, Netherlands, Belgium & EPO.
Types of Reactions Covered	Reactions of organic substances, including organometallics and biomolecules from CA Sections 20-34.
Types of Sustances Covered	Complete coverage of all substances participating in reactions including common reagents, catalysts and solvents.
Reaction Notes	Textual notes to alert users to special or unusual conditions that cannot be expressed in terms of specific substances (e.g., 'pyrolysis').
Multistep Reactions	All multistep reactions within a document will be searchable.

Figure 2. File Content for Initial Service.

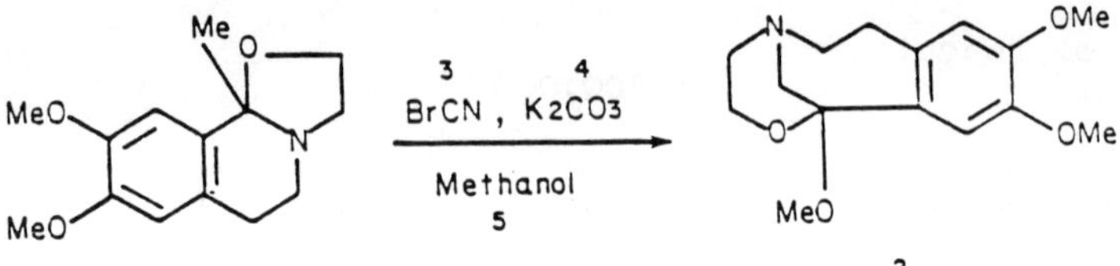

Index Entry: 1
Text Modifier: Cyanogen bromide mediated rearrangement of, methanobenzoxazonine deriv. from
Reaction Note: Product = 2, Agents = 3,4, Solvent = 5

Index Entry: 2
Text Modifier: Prepn. of

Index Entry: 3
Chemical Name: Cyanogen bromide
Text Modifier: Rearrangement of methyloxazoloisoquinoline derivs. mediated by, methanobenzoxazonines from

Index Entry: 4
Chemical Nane: Potassium carbonate
Bypass Flag: Reaction File Only

Index Entry: 5
Chemical Name: Methanol
Bypass Flag: Reaction File Only

Figure 3. Chemical Substance Indexing with Reaction Data.

Phosphorus Ylides

$$Ph_3P{=}CH{-}Ph \qquad Ph_3\overset{+}{P}{-}\overset{-}{C}H{-}Ph$$

Silatranes

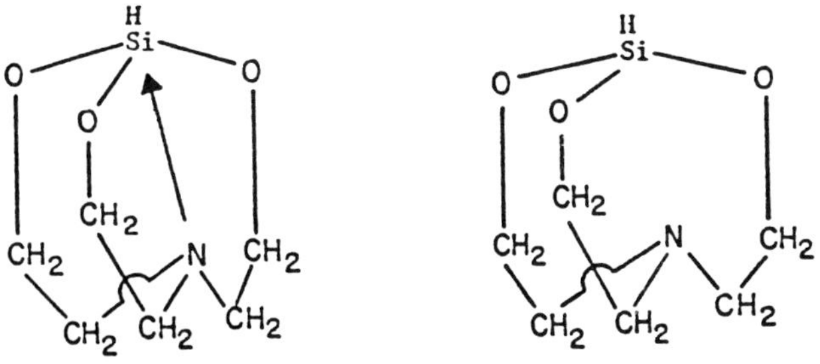

Figure 4. Examples of Compounds that are Double-Entered.

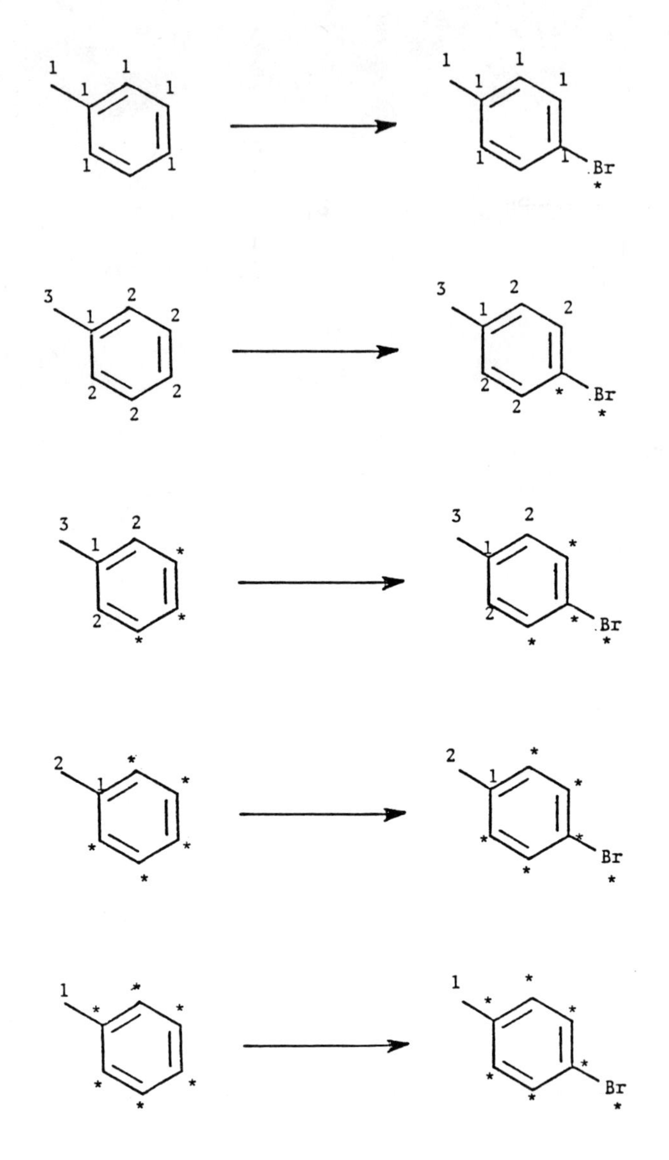

Figure 5. Attempt Match of Reactants and Products.

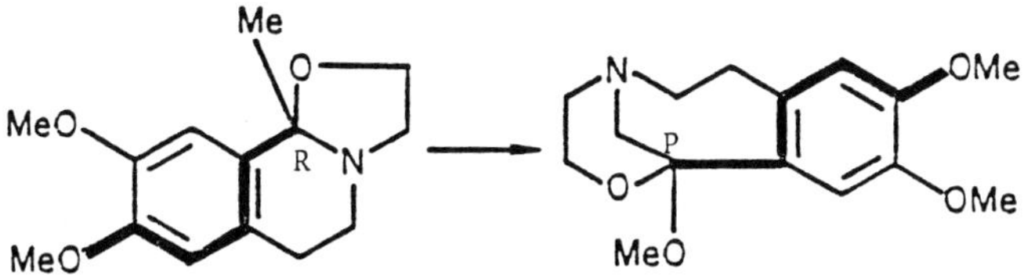

Figure 6. Location of Corresponding Reaction Sites.

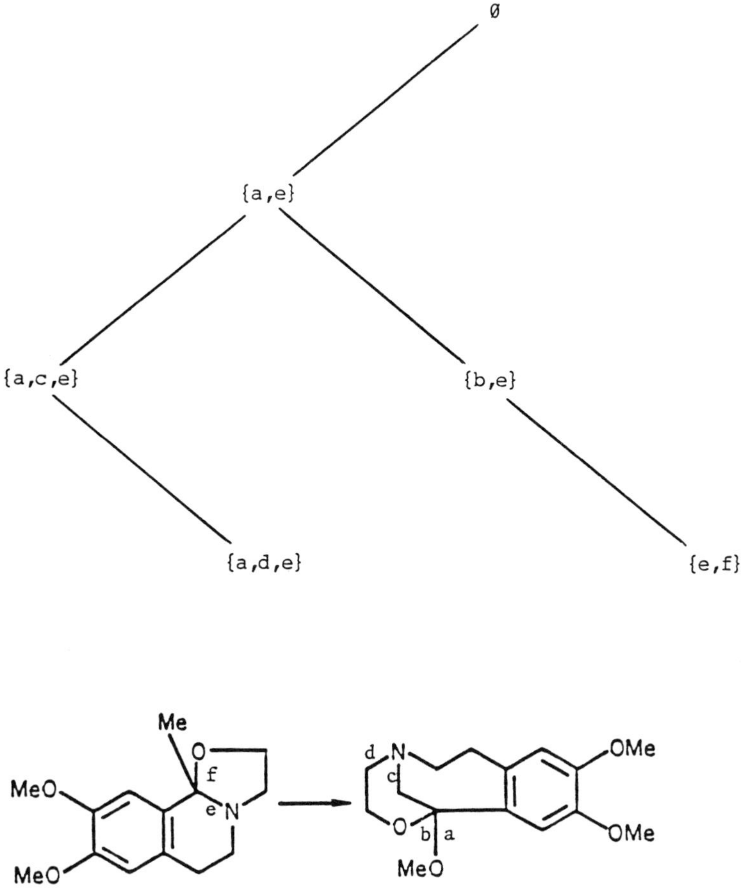

Figure 7. Tree Search of Candidate Bondsets.

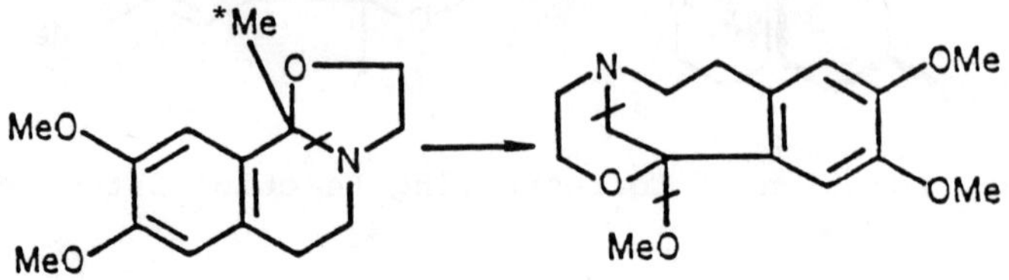

Figure 8. Reaction Site Bonds.

12 The Beilstein - Online database project

C. Jochum and S. Lawson
Beilstein - Institut

CURRENT STATE OF THE ART

The projected numerical factual database BEILSTEIN-ONLINE represents a natural extension of the BEILSTEIN Handbook of Organic Chemistry which has been published for more than 100 years in more than 300 volumes.

The present BEILSTEIN Information Pool contains Handbook and Registry data of the literature time frame from 1830 to 1960. In addition the primary literature from 1960 to 1980 has been completely abstracted. The more than 7 million abstracts consist of the structural formula, numerical physical data, reaction pathways and original literature citations. Since the handbook has been mostly published using conventional non-computerised typesetting methods this information pool is available for the most part in a printed form only. However, the most recent 25 per cent of the published volumes have been printed using electronic type-setting methods and are therefore available in a computer-readable form (Figure 1a).

This information pool will be gradually transferred into a computer-readable form and extended by additional new sources of information (see below). In the future all Beilstein information products will be generated from this information pool which will be organised in an internal database (Figure 1b).

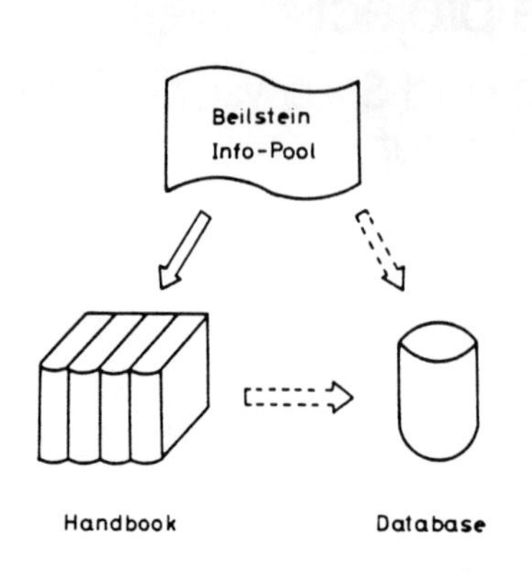

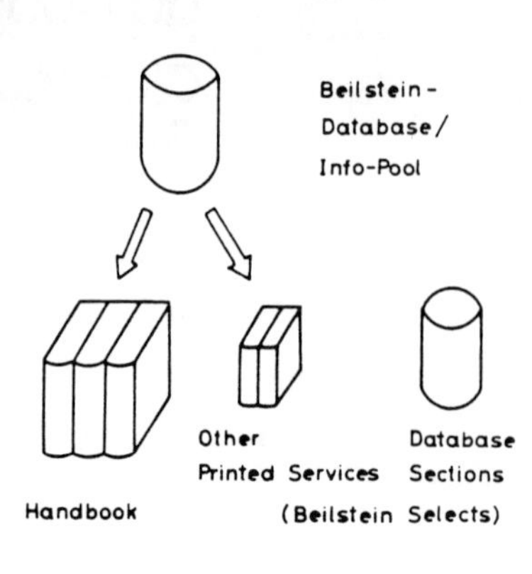

INFORMATION SOURCES OF THE DATABASE

The Beilstein database will be generated from four sources of information (Figure 2):

1) The printed handbook-series from the Basic Series up to the fourth Supplementary Series (H to E IV). These series are almost completed and contain the literature time frame from 1830 to 1960. These volumes contain the factual data of more than 1.5 million organic compounds, which makes this database worldwide unique. This represents a significant marketing advantage against other numerical and bibliographical databases.

2) The printed handbook material of the fifth Supplementary Series, which contains the literature from 1960 to 1980. As in the previous handbook series, these data have been thoroughly checked for errors and redundancies.

3) The publishing of the fifth Supplementary Series will probably continue for more than one decade. Therefore some of the excerpts will be added to the database in their original non-checked form in order to give scientists access to these data as soon as possible.

4) Beginning in 1985 the factual data of the primary literature from 1980 will be abstracted electronically. The abstracted data of the primary literature will no longer be written on paper, but entered directly in a structured manner into microcomputers and stored on magnetic diskettes. After several automatic plausibility and redundancy checks, these data will be copied on to the main frame computer and loaded into the database. Thus, the researching scientist

has in addition immediate access to the factual data of the current literature.

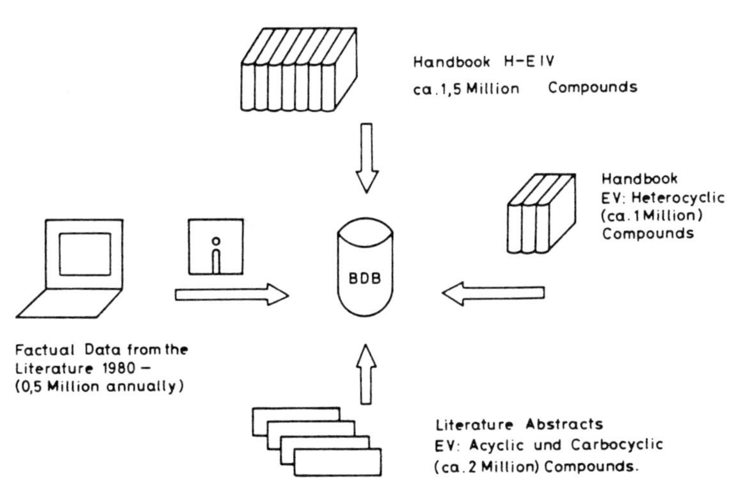

THE DATABASE CONCEPT

Taking into consideration the different sources of information, the database can be divided into a current and a refined base (Figure 3). The refined database will consist of all the data which have been published in the handbook-series and which have been checked for errors and redundancies by Beilstein information specialists. The current database is built directly from the abstracts of the primary literature without any further checking. Factual data of the current primary literature will be added to the current database on a regular basis. The refined database will be continuously extended by processing the raw material of the current database by error checking and removal of redundancies.

The data structure has been completely defined during the systems analysis phase. For each compound more than 400 different factual or numerical terms or fields can be stored. Of these 400 fields more than 300 are retrievable separately or in combination. More than 60 fields consist of numerical physical data, which can also be retrieved separately or combined with a substructure-search (see below).

Fig.3 Beilstein Database Concept

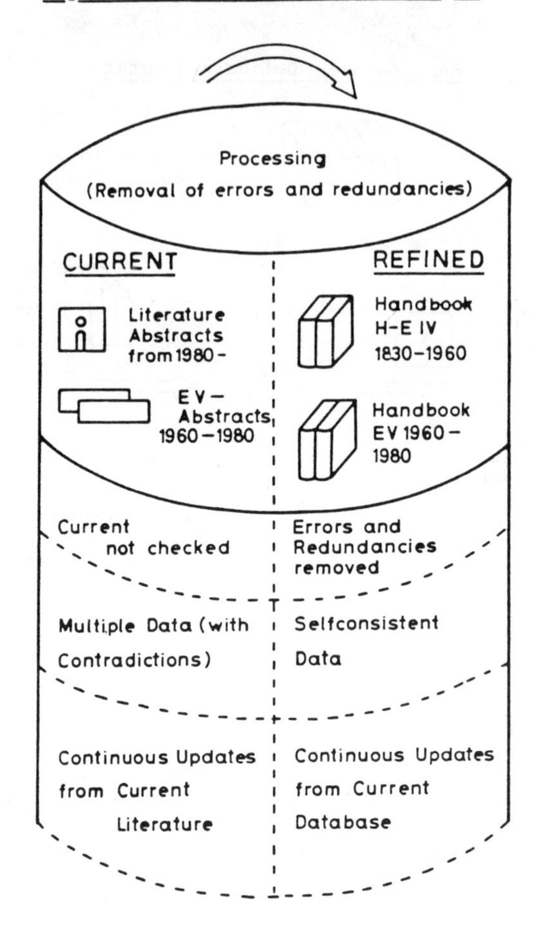

DATABASE ACCESS

The searching chemist can access the Beilstein data in several different ways:

- Via a graphical or alphanumerical input - depending on the type of computer terminal (structure or substructural search).

- By searching for numerical or alphanumerical terms (physical data, keywords, boolean terms, such as an existence of spectra, etc.). These terms can be searched separately or in boolean combinations ('and', 'or', 'not').

- By combination of the two search methods described above.

- By searching other keyfields, such as the molecular formula, the CAS registry number or the molecular structure related Beilstein Registry Number (Beilstein Lawson Number).

- The Beilstein factual database can be accessed through various Institutions and computers respectively (Figure 4).

- Via public Online-Hosts. The first Online-Host will be the Fachinformationszentrum Energie, Physik, Mathematik (FIZ-4) in Karlsruhe, which represents the German database node within the international STN-Network.

- Large chemical companies can acquire a licence for the Beilstein database to run it on their own computer centre. Thus, employees of these companies can have inhouse-access to the Beilstein data and are able to add their own data to the database. Subscribers to the Beilstein database will receive updates continuously.

- Individual customers can access selected parts of the Beilstein database by their own microcomputers, which are attached to the Winchester- or Optical Storage Device. (Beilstein-Selects).

Fig. 4: Access to BEILSTEIN-ONLINE

Fiz 4
Host

Public Online Access

BDB

Inhouse Database

Beilstein-Institut

Company XYZ

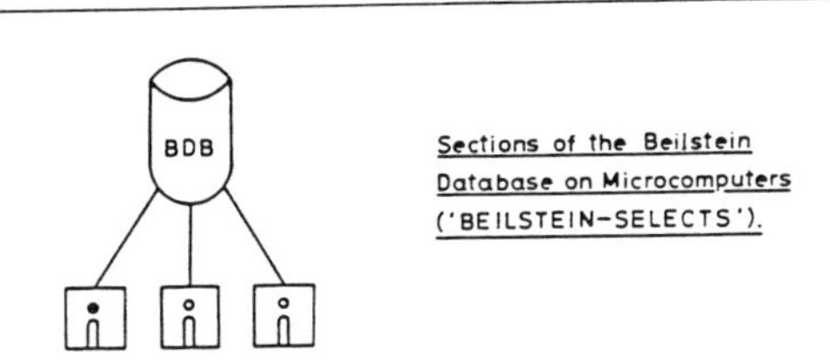

BDB
Sections of the Beilstein
Database on Microcomputers
('BEILSTEIN-SELECTS').

13 Retrieval of literature references to reactions by input of reactant and product structures

M. Bersohn
University of Toronto

The described program takes as input from the user the structure of the major organic product of a reaction together with the structure of a major organic reactant, and products in a reaction, if information about the reaction is contained in the program. The user's input is in the form of connection tables, with addenda conveying chirality and other stereochemistry.

Conventional lists of synthetic reactions in printed books use as indices the name of the type of the reaction, e.g., alkylidenation (such as in ref. 1), or the name of the product substructure, e.g., arylthioimines (2), or the element symbols of the canonically first pair of atoms between which bonds are being made and the first pair of atoms between which bonds are being broken, together with a reaction class to which the reaction can be assigned(3). These indices are useful but they do not directly address the typical user's primary problem. The typical user is a practical chemist and wants to convert a particular substance A into a particular substance B. The user wants to know if there is a reaction reported in the literature which can reasonably be expected to effect this particular transformation. If the user provides the actual structures of A and B then a computer program should be able to say which reaction retrievable by it is expected to perform the indicated transformation. In other words conventional reaction indices do not convey the scope of the reactions.

Specifying the actual structure of the molecules of interest enables the program to determine whether the stereochemistry of the reactant(s) and product are correct for the particular reaction it is retrieving. Also, functionality which can interfere with the reaction can be noticed, so that the program will provide the user only with references to reactions that are possible considering the entire molecule, not just the substructures which are changed. It seems to this writer to be unfeasible to devise an index which will not only include the produced substructure and the starting substructure but also the reaction conditions and the steric factors and the interfering groups. It follows that a computerised system that examines the entire product molecule and searches for a reaction which can produce it from a particular reactant would be much more convenient for the user than any indexed list because it would directly answer the question that he wants to ask.

The one great objection to such a system is that it requires a sizeable manpower investment for each reaction input. Matters such as reaction conditions and stereochemistry will have to be carefully examined by a knowledgeable chemist before the information is stored in the program. One might hope that this initial cost would eventually be spread over enough users so that it would be justified.

I have built a miniature version of such a system. It can be described as miniature because there are barely a thousand reactions in the reaction file. Ideally a useful file should have many thousands of reactions in it. Consider, for example, the frequently helpful two volume classic of Buehler and Pearson, the 'Survey of Organic Syntheses'.(2) It describes more than six thousand general reactions. Yet it covers heterocyclic chemistry only lightly. The synthetic parts of the huge 'Comprehensive Organic Chemistry' (4) are, collectively, several times larger than the Buehler and Pearson two volume work. A passable computer system would probably have references to 20,000 different reactions. The Derwent on-line system has more than 50,000 different reactions in it (3). It may be appropriate to comment here that my meaning of 'different' reactions is that reaction X is different from reaction Y if and only if the reactant types are different or the product types are different. Thus, for example, different ways of converting a ketone carbonyl group to a methylene are not here regarded as separate reactions. In my system they are treated internally as the same reaction with different reaction conditions and different literature references.

I first describe the system as it appears to the user. I will give a number of specific examples of the input and the resultant output. In the last part of the paper I will describe the system as it appears to the programmer.

EXAMPLES OF REPRESENTATIVE INPUT AND OUTPUT

In the table below I will present for a number of cases, the user-specified product and reactant and the reference provided by the program in response to the user's input. For the first case I provide the actual user's input, as connection tables, etc.

```
6  0  3   2  0  0  0
8  0  0   1  3  0  0
6  0  2   2  4  0  0
6  0  2   3  5  0  0
6  2  1   4  6  6  0
6  2  1   5  5  7  0
6  0  2   6  8  0  0
6  2  1   7  9  9  0
6  2  1   8  8 10  0
6  0  2   9 11  0  0
6  0  2  10 12  0  0
6  4  0  13 13 11 14
8  4  0  12 12  0  0
6  0  3  12  0  0  0
0  0  0   0  0  0  0
```

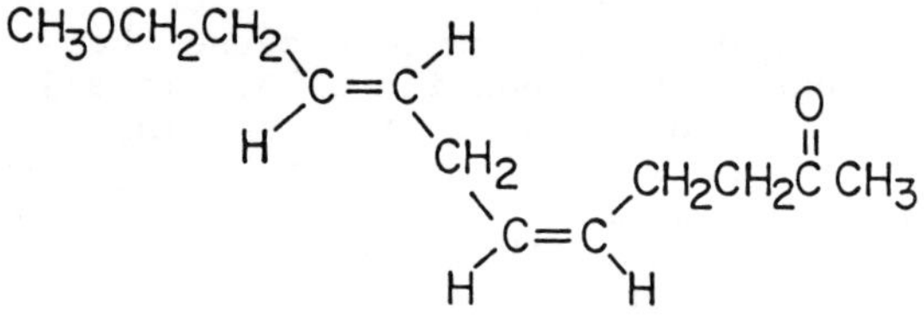

TRANS 7 4

CIS 7 10

REACTANT FOLLOWS

```
6 0 3  2  0  0  0
8 0 0  1  3  0  0
6 0 2  2  4  0  0
6 0 2  3  5  0  0
6 2 1  4  6  6  0
6 2 1  5  5  7  0
6 0 2  6  8  0  0
6 0 2  7  9  0  0
6 0 2  8 10  0  0
6 0 2  9 11  0  0
6 0 2 10 12  0  0
6 4 0 13 13 11 14
8 4 0 12 12  0  0
6 0 3 12  0  0  0
0 0 0  0  0  0  0
```

$CH_3OCH_2CH_2$

H

C=C

O

H

$(CH_2)_5$ C CH_3

TRANS 7 4

J.Lipowitz and S.A.Bowman, J. Org. Chem. 38,162(1973)

$CH_3CHOHCH_2CH_2SCH_2CH_2CH_2OH \longrightarrow$

$CH_3CHOHCH_2CH_2SCH_2CH_2CHO$

H. Tomioka, K. Takai, K. Oshima and H. Nozaki, Tet. Lett. 22, 1605 (1981)

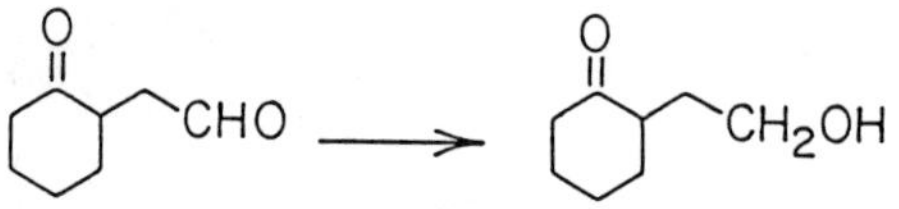

CHO ⟶ CH_2OH

Y. Maki, K. Kikuchi, H. Sugiyama and S. Seto, Tet. Lett. 263 (1977)

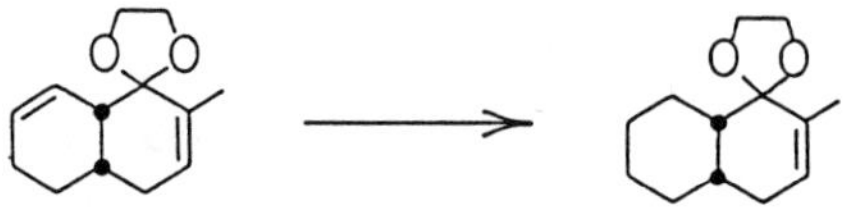

J.W. Suggs, S.D. Cox, R.W. Crabtree and J.M. Quirk, Tet. Lett. 22, 303 (1981)

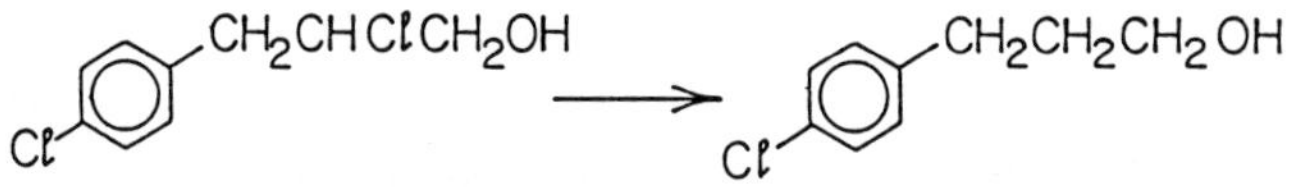

$CH_2CHClCH_2OH$ ⟶ $CH_2CH_2CH_2OH$

Cl Cl

H.C. Brown, and S. Krishnamurthy, J. Org. Chem. 34, 3918 (1969)

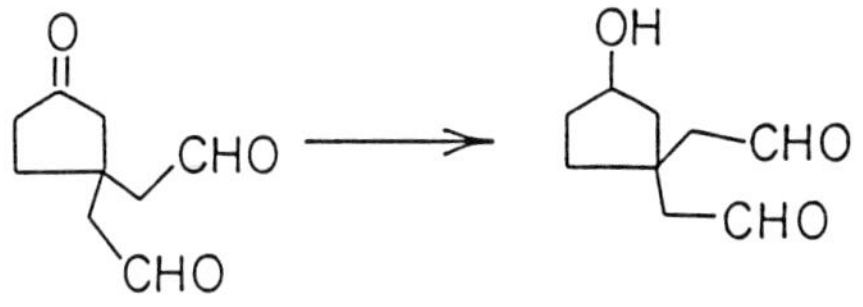

J.-L. Luche and A.L. Gemal, JACS 101, 5848 (1979)

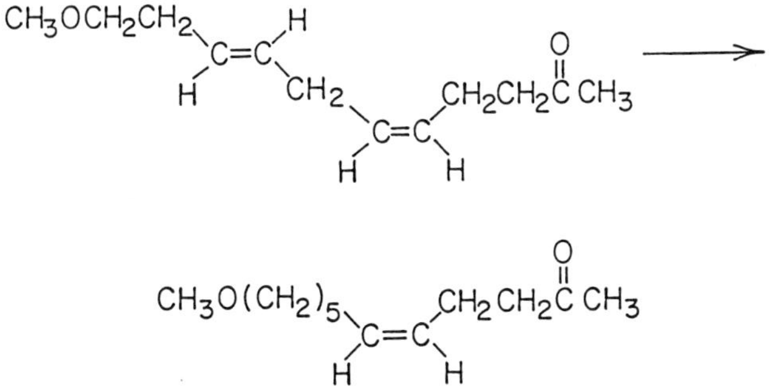

O.S. Coll Vol. V, 281 (1973)

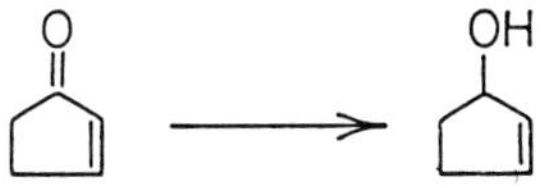

S. Krishnamurthy and H.C. Brown, J. Org. Chem. 40, 1864 (1975)

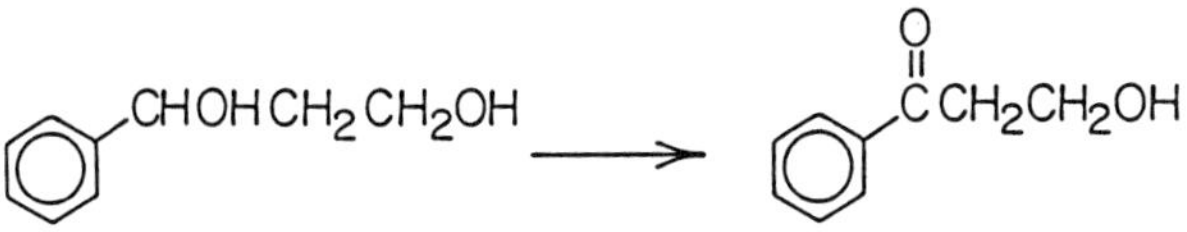

K.S. Kim, Y.K. Chang, S.K. Bae and C.S. Hahn, Synthesis, 866 (1984)

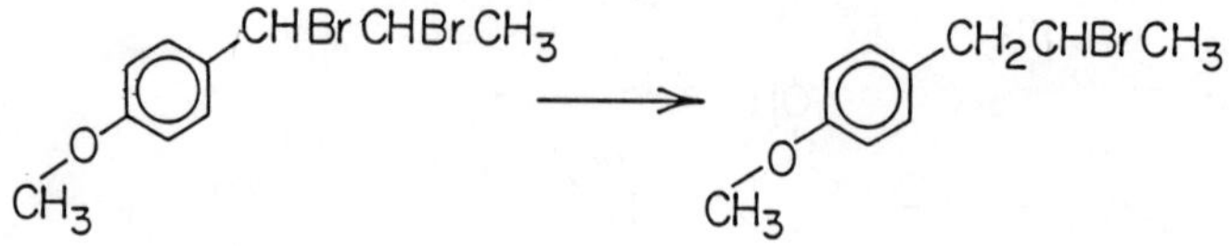

S. Kim, C.Y. Hong and S. Yang, Ang. Chem. Internat. Ed., 22, 562 (1983)

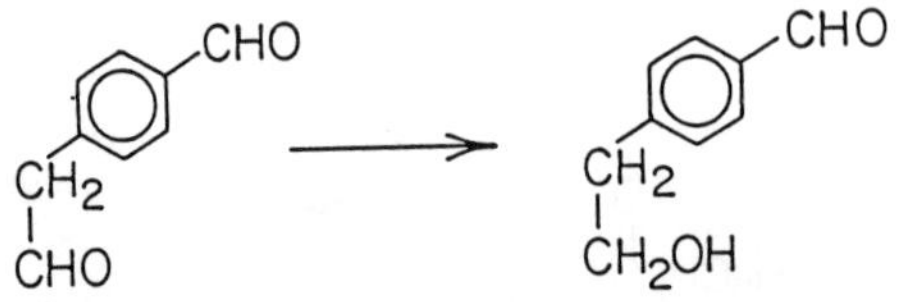

H.C. Brown, J.S. Cha, B. Nazer and N.M. Yoon, JACS 106 (8001) 1984

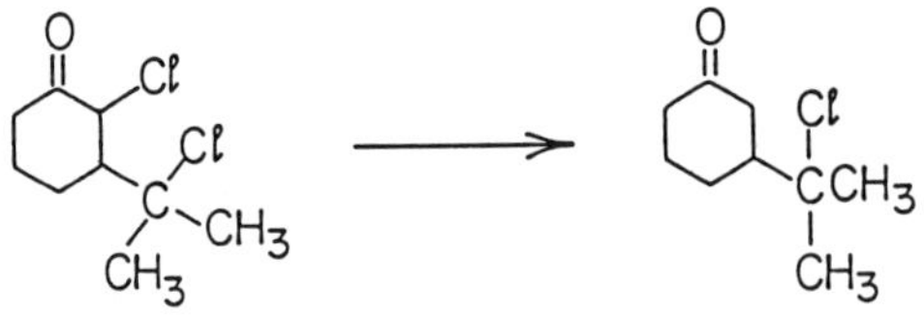

S. Takano, S. Nishizawa, M. Akiyama and K. Ogasawara, Synthesis, 949 (1984)

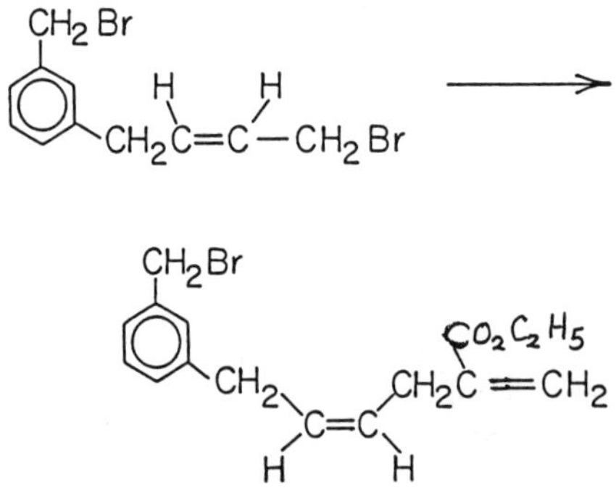

J.P. Marino and D.M. Floyd, JACS 96, 7138 (1974)

SOME DEFICIENCIES OF THE SYSTEM

1) The enforced use of connection tables cannot be said to be friendly to the user, who would prefer to draw in structures on a suitable tablet. The expense of this hardware rules it out for the present project.

2) Even with the small reaction file, usually almost a minute of IBM/3033 time elapses before an answer is obtained. With a much more sizeable file, the response time would preclude any description of this system as 'on-line'. The major reason for this is the time required for accessing the disk to retrieve that part of the program load module which contains the reaction descriptions. The reaction descriptions, other than the references, are internal to the program, in an array. However, the program, about two megabytes in size, does not fit into the available physical memory and the resultant frequent page faults account for the time consumed.

3) The system is not uniformly at the expert level in regard to what substructures are destroyed by what conditions. This means that sometimes it rejects a reaction even though it is actually feasible for the molecule concerned. Other times it suggests a reaction which the user realises will not work because a particular combination of functional groups present in the reactant(s) interferes with the reaction.

To give an example of the former error, if the user inputs the structure of the ketal I as product and the ketal II as reactant, the program cannot find any way to proceed from this reactant to the desired product. Actually, as shown by Feugeas (5), if the Grignard reagent made from the ketal II is prepared in THF, there is no difficulty in doing the indicated reaction. In ether, it is ruled out, but the program has generalised too far.

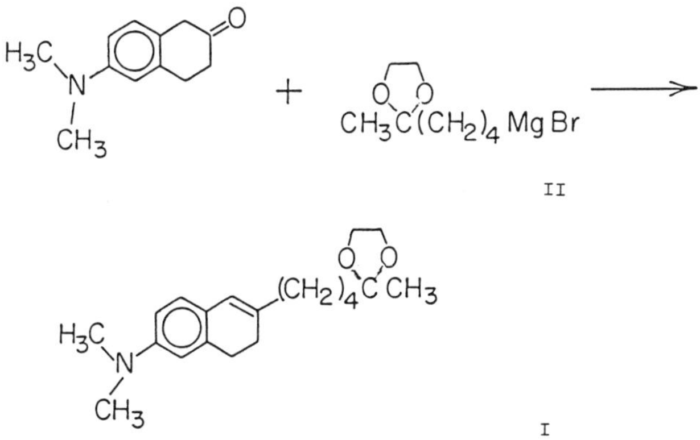

II

I

4) The system, lacking a model building facility, makes errors in deciding about steric feasibility of reactions. Consider, for example, the classic case of an α, β unsaturated ketone, in which the β carbon atom has no hydrogen atoms. In keeping with this, if it lacks equivalent ligands it becomes chiral on reduction, e.g., metal/ammonia or via hydrogenation. How are we to predict the chirality of the product, knowing the chirality of the other chiral atoms in the molecule? The program is apt to surrender and not retrieve any reference to this situation except in a few standard cases such as the production of trans 3-decalone from $\Delta\hat{4}$-3-decalone. The user will know many more specific cases and will find the program's knowledge about this far less than his.

Another example is the aldol condensation, followed by dehydration, of an aldehyde with no enolic protons and a ketone. The prediction of E and Z in the resultant dehydration to form an α, β unsaturated ketone is impossible for the program.

It is apparent that shortcomings 1 and 2 could be repaired by money, i.e., to buy graphics equipment and physical memory. Shortcomings 3 and 4 are more fundamental and future work must overcome them.

THE SYSTEM AS SEEN BY THE PROGRAMMER

Validity check on the input data

First the user's input connection table must be checked for errors.

A frequent input error is the omission of stereochemical information. The program examines each double bond to see if the substituents on either atom are equivalent. If the substituents on neither atom are equivalent and the user has not specified the double bond stereochemistry then a message is printed and the search is not performed. If the double bond is in a ring of size smaller than eight then the program assumes that the ring substituents of the vinyl carbons are cis and the user does not have to specify this fact. As for chirality, the program will tolerate one unspecified chirality if there is only one chiral atom in the molecule. If there are two or more chiral atoms in the molecule and the chirality is not completely specified then an explanatory message is printed and the search is not performed. If the user specifies chirality on a carbon which has an enolic hydrogen and is adjacent to two activating groups such as carbonyl, nitrile, sulfone, nitro etc., the program ignores the user's specification of the chirality because there is enough of the enol tautomer at equilibrium to render any specification of chirality meaningless. Internally the program does not record any chirality for any such carbon atom. The same is true for carbon atoms adjacent to vicinal dicarbonyl pairs. Finally, if there are equivalent substituted ring atoms in an achiral molecule, e.g., such as in 1,3-dibromo cyclobutane, the user must specify the cis or trans relationship of the substituents, otherwise the search is not undertaken.

In typing in the connection table the user may err in the following ways, which the program must look for.

1) No neighbours specified for an atom.

2) Atom j is included in row i as a neighbour of atom i but the number i does not appear in row j.

3) Atom j appears k times in row i but i does not appear k times in row j. A double bond is conveyed by listing the doubly bonded ligand twice.

4) The atom has the wrong number of attached hydrogen atoms. This check is applied to first row elements and silicon.

5) The number i appears in row i, a 'self-bonding' error.

6) In row i there is mention of atom j as a neighbour but row j does not exist. This is the 'nonexistent neighbour' error.

The user may sometimes correctly input a structure which cannot in practice exist. Examples are:

1) Cyclopropanone.

2) Cyclobutadiene.

3) Triple bonds or trans double bonds in rings of size smaller than eight. (I assume there is no practical demand for non-storeable molecules such as cycloheptyne.)

4) Double bonds to the bridgehead where the rings are smaller than eight.

5) Cyclopentadiene structures such as 1-alkyl,2,4-cyclopentadiene, where the double bonds could not stably remain.

6) Cyclohexadienyl substructures which have a third, exocyclic double bond, and which can aromatise.

7) Molecules, such as 1-amino-2-chloroethane, which cannot be stored because the amino nitrogen will displace a leaving group within reach in the same molecule.

The program must recognise these cases and reject them at input time, whether they are proposed products or proposed reactants. Most of the time these errors arise from the user's carelessness, which should not be underestimated.

I have even installed a check which causes the program to exit without retrieval attempt if the reactant and the product are the same.

Canonicalisation of the structure representation and recognition of reactants in synthetic reactions

Usually this is not a problem since the connection table produced for the reactant by the prescribed manipulations of the connection table of the product, when canonicalised, is identical with the canonicalised version of the user's connection table. However, recognition of tautomers has proved to be a problem. The user may input, for example, a cyclic hemiacetal, but the program may generate the structure of the corresponding open chain aldehyde alcohol. The way around this difficulty that I have adopted is to direct the cyclic hemiacetal and cyclic hemiketal cases to a specific routine which constructs an alternative connection table. This alternative structure representation is used for special purposes such as comparison with other molecules. Certain phenols are significantly tautomeric with ketones and this situation has not yet been included in this program. To facilitate substructure recognition the connection tables are canonicalised, in a way that has been previously described. After this is done the rings are found and then the various substructures which might be produced by reactions present in the file are collected, basically as described ten years ago (6,7,8,9). The key result is a set of numbers referring to the substructures discovered in the product molecule which is known to be produced by a reaction in the file. These numbers are used as indices to retrieve the reaction descriptions from the file. In addition it is convenient to establish a table to display for each atom in the product molecule the number of the corresponding atom, if any, in the user-specified reactant molecule. Occasionally, when there are two significant organic reactants, since the user has only specified one, ambiguities arise and such a table cannot usefully be established. In such a situation there is a greater average expenditure of CPU time in locating the proper reaction.

As an example, let us assume the user has input the structure of ethyl 4-hydroxybutanoate, $HO(CH_2)_3CO_2C_2H_5$, as the desired product, and malonic ester as the reactant. The program finds seven correspondences as shown in the figure below.

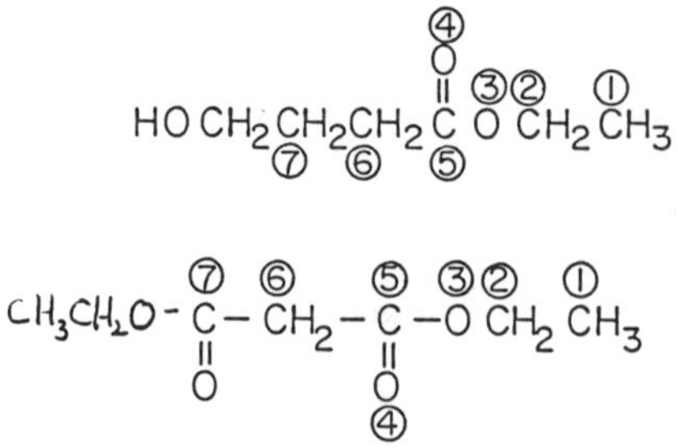

④ ③ ② ①
$HO\,CH_2CH_2CH_2C(=O)\,O\,CH_2\,CH_3$
⑦ ⑥ ⑤

⑦ ⑥ ⑤ ③ ② ①
$CH_3CH_2O-C(=O)-CH_2-C(=O)-O\,CH_2\,CH_3$
④

However, these correspondences preclude the true reaction, of malonic ester and oxiran, since the user did not specify oxiran as a reactant and he did not specify carbon dioxide and ethanol as products of the reaction. Therefore, after an unsuccessful search for a possible reaction, given this assumed correspondence of product atoms and reactant atoms, the program begins again without assuming any correspondence. On this second try, various reactions are found by the program which can give rise to ethyl 4-hydroxbutanoate in one step, but only one of these is found to start with malonic ester, and it is therefore only the reference for this one which is written out to the user.

Evidently we could shorten the response time by forcing the user to give all reactants and all products but I assume that few users would favour this.

The information about a reaction appears in two places in this system. The literature reference, i.e., date, journal, page, author(s) are contained in a file external to the program. The rest of the information about the reaction is called the reaction description and each such description is an element in an array internal to the program. Reaction descriptions contain the usual indications to replace a bond from i to j with a bond from i to k, etc. (10).

After the reaction is retrieved the program must perform a few associated tests to make sure that the current molecule is actually a suitable product of the reaction, taking into account the given reactant. We have to answer the following questions.

1) Are there interfering groups present?

2) Is the exact substructure present in this molecule produced by the reaction?

For example, if the reaction is alkylation of a β-keto ester there should be an alkylating carbon, i.e., the α carbon atom between the two carbonyl groups should have at least three carbon atom neighbours.

3) Is there too much steric hindrance?

This matter is ascertained in the crudest fashion, e.g., by counting the branches on α and β atoms, or, where appropriate, considering the cis adjacent substituents in a ring of size smaller than six and the 1,3-diaxial interactions in rings of size six.

4) Does the reaction make bonds which are already present in the indicated reactant molecule?

If so, the program does not consider this reaction any further. The bonds present in the reactant molecule and still present in the product molecule are inferred from the correspondence previously found between reactant atoms and product atoms.

5) Does the reaction make all the bonds which are necessary to attach the 'extra pieces' to the reactant?

The atoms of the input product molecule can be assigned to two classes. In one class there is a corresponding atom in the starting molecule. In the other class there is no such corresponding atom. All the bonds between atoms of the first class and atoms of the second class must be made by the reaction or by repetition of the reaction. If the reaction does not have this property then it should not be used. Application of this test saves much CPU time.

When the reaction is thus verified as appropriate the subroutines are called which generate the connection tables of the reactant(s) from that of the product. The reactant(s) is compared with the reactant requested by the user. If it coincides then the references are fetched from an external file, indexed by the reaction number. The array of reactions internal to the program consist of instructions for performing detailed tests on the produced substructure, bond replacements, stereochemical changes, etc. (10)

Versions of this system exist in PL/I and IBM assembly language.

It may generally be observed that synthetic organic chemists with a Ph.D are not fond of computers. The inhuman speed of the computer and its lack of human forgetfulness must be accompanied by an inhuman omniscience or else such a user is inclined to dissatisfaction. Aside from these problems with potential users, drawbacks of the system itself, requiring the input of connection tables and frequently consuming an expensive minute of highest priority CPU time, also suggest that any widespread use of the system described here is in the distant future.

REFERENCES

1. J. Mathieu and J. Weill-Raynal, 'Formation of C-C bonds', Volumes I, II and III, Georg Thieme Verlag, 1979.

2. C.A. Buehler and D.E. Pearson, 'Survey of organic syntheses', John Wiley & Sons, Inc., New York, volume 1: 1970, volume 2: 1977.

3. Theilheimers's synthetic methods of organic chemistry, S. Karger, Basle, A.F. Finch ed., (annual yearbooks) 1984.

4. D. Barton and W.D. Ollis, eds, 'Comprehensive organic chemistry', Pergamon Press, Oxford, 1979.

5. C. Feugeas, Bull. Soc. Chim. France 2568, 1963.

6. A. Esack and M. Bersohn, J. Chem. Soc. Perkin I, 1124, 1975.

7. M. Bersohn and A. Esack, Chemica Scripta, 211, 1976.

8. A. Esack, J. Chem. Soc. Perkin I, 1120, 1975.

9. M. Bersohn, J. Chem. Soc. Perkin I, 1239, 1973.

10. M. Bersohn and A. Esack. Computers & Chemistry, 1, 103, 1976.

14

Automatic keyword generation for reaction searching

A. P. Johnson and A. P. Cook
University of Leeds

ABSTRACT

There has been a recent trend towards a purely graphical approach to reaction retrieval systems, with virtual neglect of keyword terms. However, keywords can play an important role in reaction indexing, as they can usefully convey reaction concepts which are difficult or impossible to formulate purely in terms of structural features. It is felt that the implementation of an 'expert assistant', to aid the selection of appropriate keywords at registration and to help formulate queries, will allow keywords to be used in a reliable and user-friendly environment.

This paper presents an overview of the current implementation of keyword generation and keyword searching in ORAC (1). The modules required for the automatic recognition of reaction types, and hence the automatic selection of keywords, are identified. Finally, an improved design of a keyword thesaurus is discussed.

INTRODUCTION

In the design of ORAC, emphasis has been placed on maximising the use of graphics for both data input and for query generation. The use of menus throughout the program, to aid rapid data registration and query generation, was also deemed desirable. This approach is beneficial to the casual user, as it provides both a natural language interface (drawing of structures) and organises the sequences of actions the user has to perform to formulate a query. The inclusion of an 'on-line' help facility also aids the interactive use of the program.

However, blanket adherence to the above requirements is not always desirable since not all the information necessary to query a chemical reaction can be conveyed in purely pictorial form or be easily selected from menus. Many descriptions of overall reaction features such as mechanism, selectivity or reagent types are better expressed in the form of keywords. Hence, from the very beginning, keywords have formed an essential part of the description of a reaction entry in ORAC.

OVERVIEW OF ORAC

Before discussing the main topic of this paper, a brief overview of ORAC will be presented from two viewpoints: registration of new reactions and search of the database.

If the user has the appropriate privilege, then selection of the 'ADD A NEW REACTION' button on ORAC's startup page leads to the display of a sketchpad (Figure 1). All the reactants are now drawn (up to a maximum of three individual compounds) in the sketch window and are designated as such by selecting the 'REACTANT' button on the right hand side of the page. The procedure is repeated for the reaction products (maximum of two) but designated with the 'PRODUCT' button. The progress of the reaction scheme is shown in a small status window at the bottom right of the sketch page (Figure 2). When the user is satisfied that the scheme is complete, then the 'ENTER' button is selected.

After the system has indexed the information regarding the reaction change (Figure 3), the user is taken to an empty KEYS page (Figure 4). All items that are entered on this page are searchable in some form or another. The user now fills the journal, year, page, yield, temperature and author fields with the relevant data that have been abstracted from the source reference. Some additional data fields that appear on this page are headed 'REACTION KEYS', 'REAGENT KEYS' and 'SOLVENT KEYS'. For these, users are expected to use their chemical expertise to select appropriate words that describe the general features of the current reaction. Clear-cut selection rules and on-line access to libraries of existing keywords are available as an aid to the novice user. Thus, if a user tries a particular keyword but finds that it is not recognised by the system, it is possible to scan the keyword thesaurus for suitable candidates (Figure 5). If none is applicable, then new words can be readily added to the libraries, either as discrete words or as synonyms of existing words.

When all data items have been added to the keys page, the text fields on the data page (Figure 6) are filled. Finally, activation of the 'FILE' button initiates database insertion.

In search mode, the user can use the option buttons 'REACTION KEY', 'REAGENT KEY' and 'SOLVENT KEY' to gain access to the keyword libraries. This is done by selecting one of the buttons and then replying to the prompt with a question mark. The library is now available for alphabetical searching from any specified point. Once the user has found a suitable keyword, it can then be added to the current search query. For example a legal reaction keyword is 'STEREOSELECTIVE' and this may be added to an existing query with the 'AND' operator to ensure that the retrieved reactions conform to this requirement.

Our experience of constructing the database over the last twelve months, has allowed us to identify some advantages and disadvantages of our present system which are listed below:

1. Keywords convey concepts not easily represented in graphical notation. Examples can be found in queries such as:

   a) 'Find reactions of type X that are mediated by ACID CATALYSIS'

   b) 'Find reactions of type X that are not conducted in APROTIC solvents'

   c) 'Find all REDUCTIONs of ketones that are STEREOSELECTIVE'

2. The encoder sometimes finds it difficult to make the optimum choice of keywords from the set of possible candidates. A poor choice can lead to inconsistent use of keywords and reduces the reliability of using keywords as a search tool. In addition, this dilemma can slow down the encoding of a new reaction. The encoder must have a good understanding of the chemistry involved to make the appropriate selections and must also have some knowledge of the contents of the keyword thesaurus to be productive.

3. The only relationship that currently exists between keywords is the possibility of equivalence (synonyms). Otherwise each word is a discrete entity. This leads to problems with certain searches such as:

   'Find all ACYLATION reactions'

   If the encoder has chosen to use a keyword such as FORMYLATION or ACETYLATION in preference to ACYLATION, then total recall is impaired as both FORMYLATION and ACETYLATION are members of the set of ACYLATION reactions and should be included in the search results. The present solution is to register both the words ACYLATION and FORMYLATION at the expense of using up an extra keyword field. Of course, this places extra demands on the encoder in terms of recognising relationships between keywords and determining the applicability of such relationships as a search aid for each reaction.

As a consequence of the last two points, the goals of the proposed new system are as follows:

1. To ensure consistency in the use of keywords, the system itself must do its utmost to generate sets of keywords derived from perceived structural and reaction features. For keywords that cannot be derived directly from the reaction scheme or reaction conditions, the system must assist the encoder in choosing suitable keywords from predefined lists. This entails a screening system which prunes the possible sets of keywords to a relevant list for inspection.

2. A simple semantic network of key words and key phrases must be implemented and rules must be developed for keyword interpretation at registration and search times.

REACTION CHARACTERISTICS INDEXED BY ORAC

Version 4 of ORAC allows the following searchable data items to be associated with each reaction record:

Author names	(text, 5 fields)
Journal, page and year	(text, 1 field)
Reactants	(connection table, 3 fields)
Products	(connection table, 2 fields)
Reagent names	(text, 5 fields)
Reaction name	(text, 1 field)
Reaction keywords	(text, 5 fields)
Reagent keywords	(text, 5 fields)
Solvent keywords	(text, 3 fields)
Temperature	(numeric, 1 field)
Yield	(numeric, 1 field)

In future versions of ORAC this list will be expanded to include:

Intermediates	(connection table, 2 fields)
By-products	(connection table, 2 fields)
Synthon	(connection table, 1 field)

It is also intended that reagents should be searchable by structure and formula as well as by name, and that the fixed fields for keywords be converted to unlimited lists.

From the above explicitly provided data, ORAC derives a reaction map (i.e. a list of reactant-atom to product-atom correspondences) that characterises the particular reaction. This procedure is semi-automatic, as the encoder is required to check and possibly edit the map after application of ORAC's mapping algorithms.

Keywords are currently supplied by the encoder to describe reaction, reagent and solvent properties. In the future versions of ORAC this list will be sub-divided and extended (Table 1), as the three current categories have been found to be too restrictive.

Table 1.

KEYWORD TOPIC	SCOPE	EXAMPLES
Reaction	Selectivity, control, topological or geometric changes	REGIOSELECTIVE, STEREOSPECIFIC, CYCLISATION, RETENTION.
Intermediate	Reaction intermediates	CARBONIUM ION, ENOLATE, RADICAL, CARBENE.
Mechanism	Probable reaction mechanism	SN1, E2, SN2', RADICAL COUPLING.
Reactant	Reacting functionality	ENOL, KETONE, THIOL.
Product	Reaction goals, new functionality	ALLYLIC ALCOHOL, ALKYL CHLORIDE, ENONE.
Reagent	Reagent properties	AMINE BASE, LEWIS ACID, CATALYST.
Solvent	Solvent properties	APROTIC, DIPOLAR, NON-POLAR.

MODEL FOR AUTOMATIC KEYWORD GENERATION

From the above description of the reaction model employed by ORAC, it can be seen that some classes of keyword could be inferred from the set of structures, reaction conditions and the reaction map. In contrast, features such as reaction selectivity may be difficult to infer directly from an individual reaction entry. In this situation the program should suggest candidates for manual selection by the encoder.

The goal of a keyword generator is to identify the type of reaction that has in fact taken place. This is in contrast to systems such as LHASA (2) and SECS (3) which employ production rules (transforms) to generate possible precursors to a target molecule. In essence ORAC must apply production rules in a recognitive, as opposed to a generative sense. To associate a given reaction with plausible reaction descriptors, two major components need to be defined. They are a reaction map generator and a map tester. The overall scheme indicating the relationship of these two modules is shown in Figure 7.

a) Map generator

The implementation of mapping algorithms for the detection

of ostensible reaction bond changes has been discussed elsewhere (4,5). The generator produces a set of alternative solutions (maps) to the problem of mapping the reactant graph on to the product graph. Each map has a corresponding rating value associated with it, based on the number and types of altered bonds. The smaller the number of bonds made, broken or altered, then the higher the map is ranked as 'good'. The maps are sorted on the basis of the rating before being tested. The rating heuristic follows from the simple observation that most reactions involve a minimal amount of bond change.

b) Map tester

Each map produced by the Generator is matched against reaction descriptors in a database. Successful matches invoke the application of production rules to select candidate keywords. Unsuccessful matching reinvokes the Generator to supply the next best map. The testing process is repeated, with failures reinvoking the Generator until no more maps of a useful rating are produced. At this point the encoder manually constructs the map for testing. Any failure now, to identify the reaction, constitutes the discovery of a new reaction as far as the reaction knowledge base is concerned.

An essential requirement of the Map Testing module is the interaction with a knowledge base containing descriptions of reaction core substructures (6), i.e. in the reactant and product, those portions of the graph which are changed by the reaction. These substructures provide a compact reaction description through inclusion of all the bond changes brought about by the reaction. Each substructure combination is associated with particular keywords. Work is currently in progress on the definition and implementation of a grammar to describe structural fragments involved in reactions (6,7). The grammar must be able to define simple generic features such as the pattern of bond changes needed to identify reaction classes, for example substitution, elimination, addition or insertion. It must also be able to identify specific examples of certain reaction classes. This is illustrated in Figure 8. The reaction notation is derived from a notation due to G. Vladutz (9).

AN IMPROVED KEYWORD THESAURUS

A simple scheme to represent keywords is now outlined. The scheme presented is by no means complete, but serves to illustrate the strategies that could be used in registration and search interpretations of keywords (8). The keyword scheme is composed of RELATIONSHIPS and RULES.

Relationships are used to represent explicit facts about pairs of keywords. These take the form:

P(X, Y).

where X and Y are keywords and P denotes a relationship. It should be read as X is related by P to Y.

Using relationships such as 'synonym_of' (to denote exact equivalence) and 'includes' (to denote that one word encompasses the concept of another word, i.e. a hierarchy), then examples from our keyword libraries expressed in this form are listed in Table 2.

Table 2.

a) Equivalences

```
synonym_of (RING CLOSING, RING FORMING).
synonym_of (RING FORMATION, RING FORMING).
synonym_of (LACTONISE, LACTONISATION).
```

b) Concepts

```
includes (RING FORMING, LACTONISATION).
includes (RING FORMING, CYCLOADDITION).
includes (CYCLOADDITION, DIELS ALDER).
includes (SUBSTITUTION, NUCLEOPHILIC SUBSTITUTION).
includes (SUBSTITUTION, ELECTROPHILIC SUBSTITUTION).
includes (NUCLEOPHILIC SUBSTITUTION, SN2).
includes (NUCLEOPHILIC SUBSTITUTION, SN2').
includes (NUCLEOPHILIC SUBSTITUTION, SN1).
includes (ACYLATION, FORMYLATION).
includes (ACYLATION, ACETYLATION).
```

Rules allow inferences to be made about keywords. Keywords need to be interpreted at both registration and query time, as shall now be illustrated.

A sample set of production rules are outlined in Table 3. These operate on the example relationships given in Table 2. These production rules are used to define the processing rules of keywords at registration and query time (Table 4).

The Exchange rule simply states that if a registered keyword is synonymous with other words, then choose one of these words to represent the set. At search time, if any other member of this set has been chosen for the query, then it must be translated to the representative word as only this one is held in the reaction database.

The Redundancy and Expansion rules are complementary. The first removes redundant keywords at registration, while the second expands the meaning of a query keyword to include sub definitions. This is illustrated with an example query:

'Find all ACYLATION reactions'

This would be translated by application of the Expansion rule into:

'Find all ACYLATION or FORMYLATION or ACETYLATION reactions'

The Redundancy rule has ensured that no instance of a reaction being described as both ACYLATION and FORMYLATION or as both ACYLATION and ACETYLATION has occurred, hence the three keyword components of the final query will not produce redundant database matches.

Table 3.

a) Exchange Rule.

If, in a list containing keywords, any word X can be proved to be synonymous (using 'synonym_of') with another word Y (not necessarily in this list), then exchange X for Y.

b) Duplicate Rule.

If, in a list containing keywords, a word X occurs more than once, then remove the duplicates.

c) Concept Redundancy Rule.

If, in a REGISTRATION list, there are two words X, Y such that it can be proved that X encloses the concept of Y (using 'includes'), then X is removed.

d) Concept Expansion Rule.

If, in a QUERY list, there is a keyword X that can be proved to enclose the concepts of words A, B, C ... (using 'includes'), then replace X with the expression [X or A or B or C ...].

Table 4.

a) Registration Processing Rule (in part).

For each keyword that occurs in the list of registration data, apply the following rules:
  Exchange rule ;
  Duplicate rule ;
  Concept Redundancy rule ;
  ...

b) Query Processing Rule (in part).

For each keyword that occurs in the query list apply the following rules:
  Exchange rule ;
  Concept Expansion rule ;
  ..

The above rules illustrate how the system could intelligently screen keywords at registration, and apply rewrite rules to the search query. Rewrite rules are used to both interpret the implied meaning of the query and reorder the elements of the query for maximum search efficiency. The latter application has not been discussed here.

SUMMARY

To improve the implementation of keyword searches in ORAC, the solution proposed is to develop a knowledge base composed of rules relating structural and reaction features to lists of candidate keywords. This will allow an 'expert assistant', having mapped a reactant graph to a product graph, to recognise when a keyword could be associated with a given reaction. This knowledge base must also contain a hierarchical keyword thesaurus with associated production rules to allow transformation of database queries into the correct form for full search recall.

REFERENCES

1) ORAC is an acronym for Organic Reactions Accessed by Computer. This is a reaction database management system designed and developed at the University of Leeds under the direction of A. P. Johnson.

2) E. J. Corey, A. K. Long and S. D. Rubensstein, Science, 228, 408, 1985.

3) W. T. Wipke, G. I. Ouchi and S. Krishnan, Artif. Intell., 11, 173, 1978.

4) C. Marshall, 'Computer Assisted Design of Organic Synthesis', thesis, University of Leeds, 1984.

5) a) J. J. McGregor and P. Willett, J. Chem. Inf. Comput. Sci., 21, 137, 1981.

   b) M. F. Lynch and P. Willett, J. Chem. Inf. Comput. Sci. 18, 154, 1978.

6) The use of chemical grammars in the LHASA pattern chemistry executive is described by G. A. Hopkinson, 'Computer-Assisted Organic Synthesis Design', thesis, University of Leeds, 1985.

7) Grammars have been used in chemistry in both recognitive and regenerative senses in such areas as generic chemical structure searching and synthesis planning. The reader is referred to the following paper and the references cited within:

   S. M. Welford, M. F. Lynch and J. M. Barnard, 'Chemical Grammars and their Role in the Manipulation of Chemical

Structures', J. Chem. Inf. Comput. Sci., 21, 161, 1981.

8) A model thesaurus, implementing the discussed production rules and keyword relationships, has been written in Prolog under the Sussex Poplog environment.

9) G. Vladutz, paper in this volume.

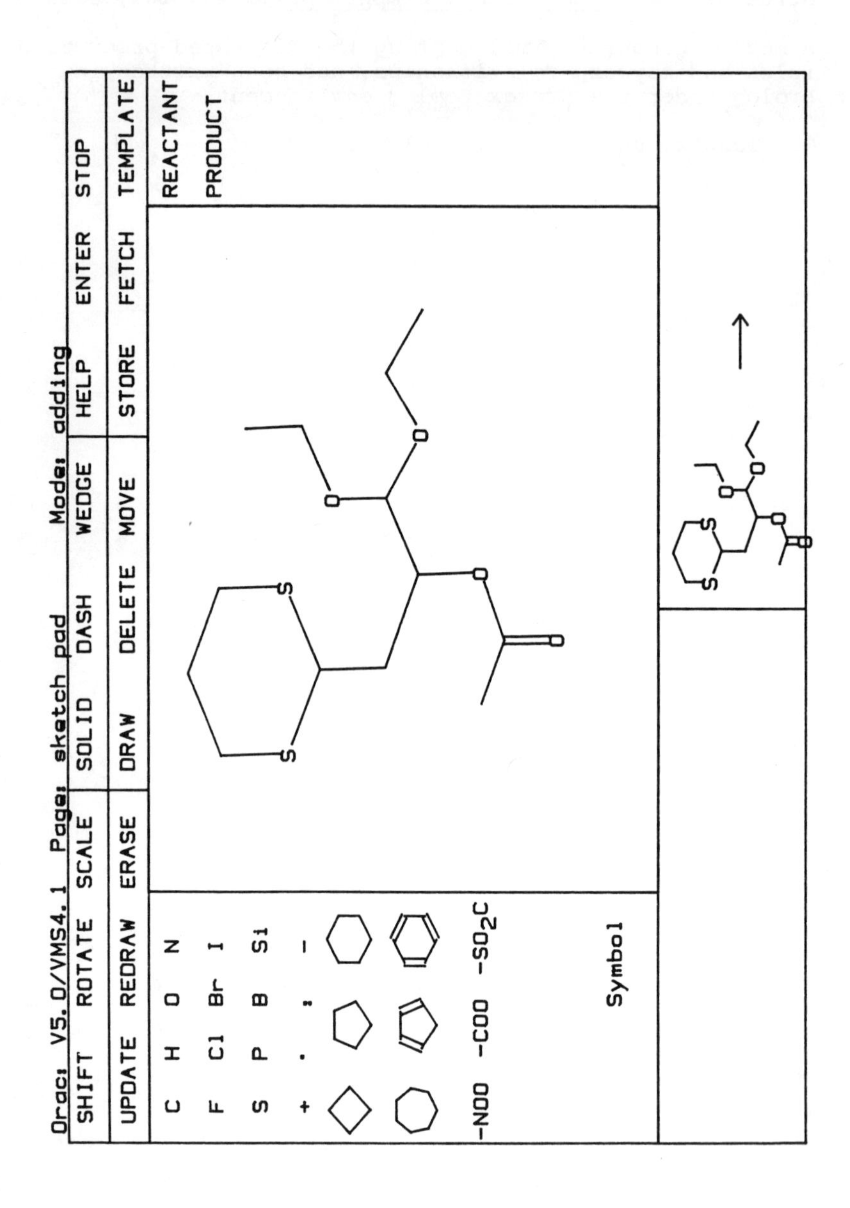

Orac: V5.0/VMS4.1 Page: sketch pad Mode: adding
SHIFT ROTATE SCALE SOLID DASH WEDGE HELP ENTER STOP
UPDATE REDRAW ERASE DRAW DELETE MOVE STORE FETCH TEMPLATE
REACTANT
PRODUCT
C H O N
F Cl Br I
S P B Si
+ . : -
-NOO -COO -SO2C
Symbol
S
S
O
O
O
O

Figure 1.

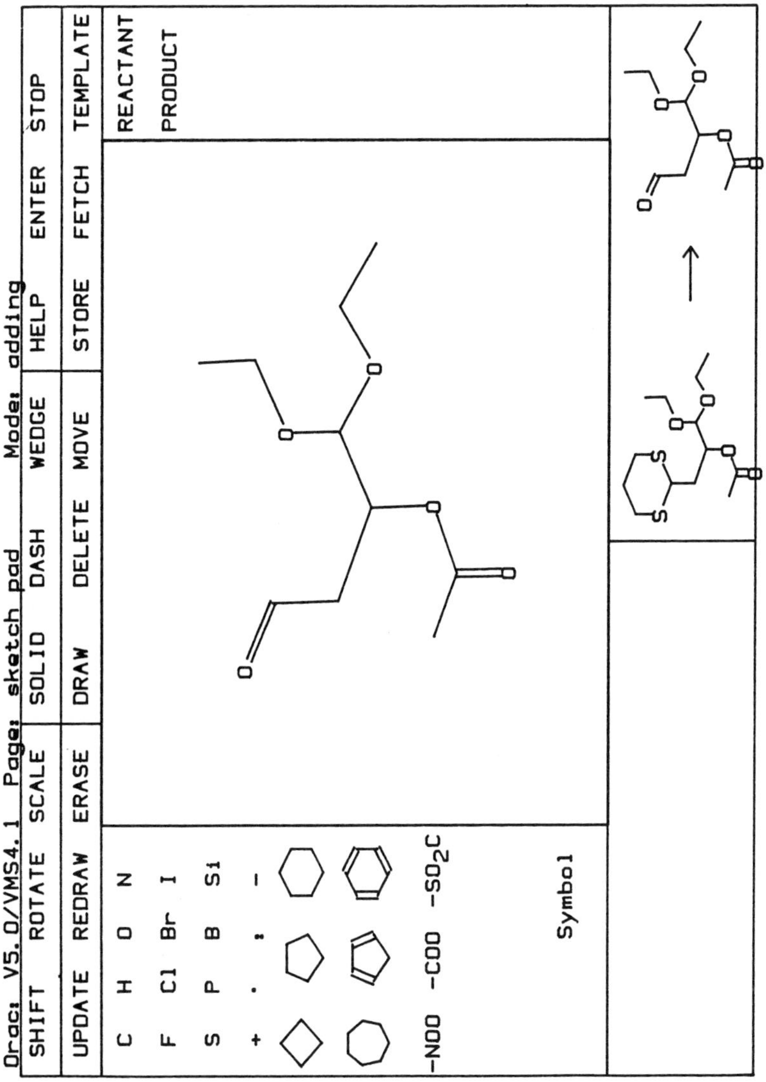

Orac: V5.0/VMS4.1 Page: sketch pad Mode: adding
SHIFT ROTATE SCALE SOLID DASH WEDGE HELP ENTER STOP
UPDATE REDRAW ERASE DRAW DELETE MOVE STORE FETCH TEMPLATE
REACTANT
PRODUCT
C H O N
F Cl Br I
S P B Si
-NO2 -COO -SO2C
Symbol
O
S
S

Figure 2.

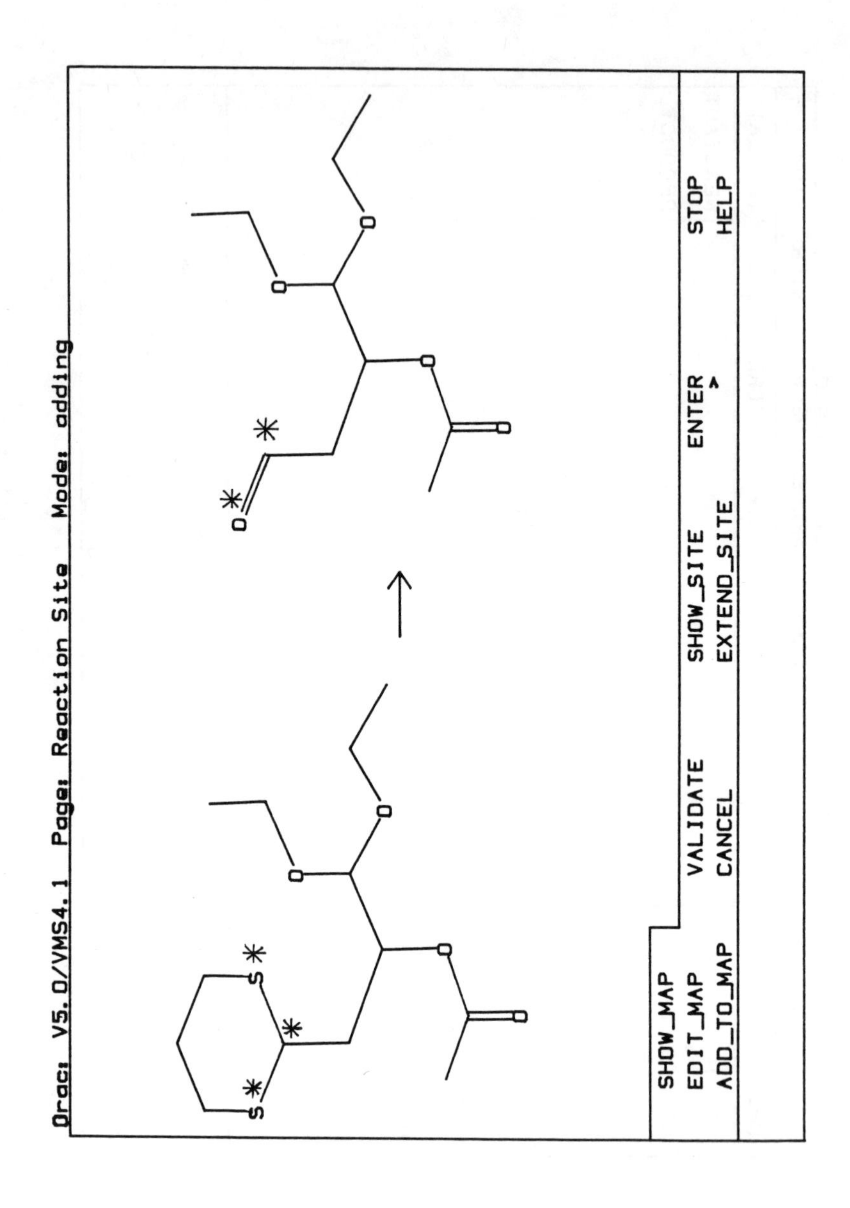
Orac: V5.0/VMS4.1 Page: Reaction Site Mode: adding
SHOW_MAP
EDIT_MAP
ADD_TO_MAP
VALIDATE
CANCEL
SHOW_SITE
EXTEND_SITE
ENTER
STOP
HELP

Figure 3.

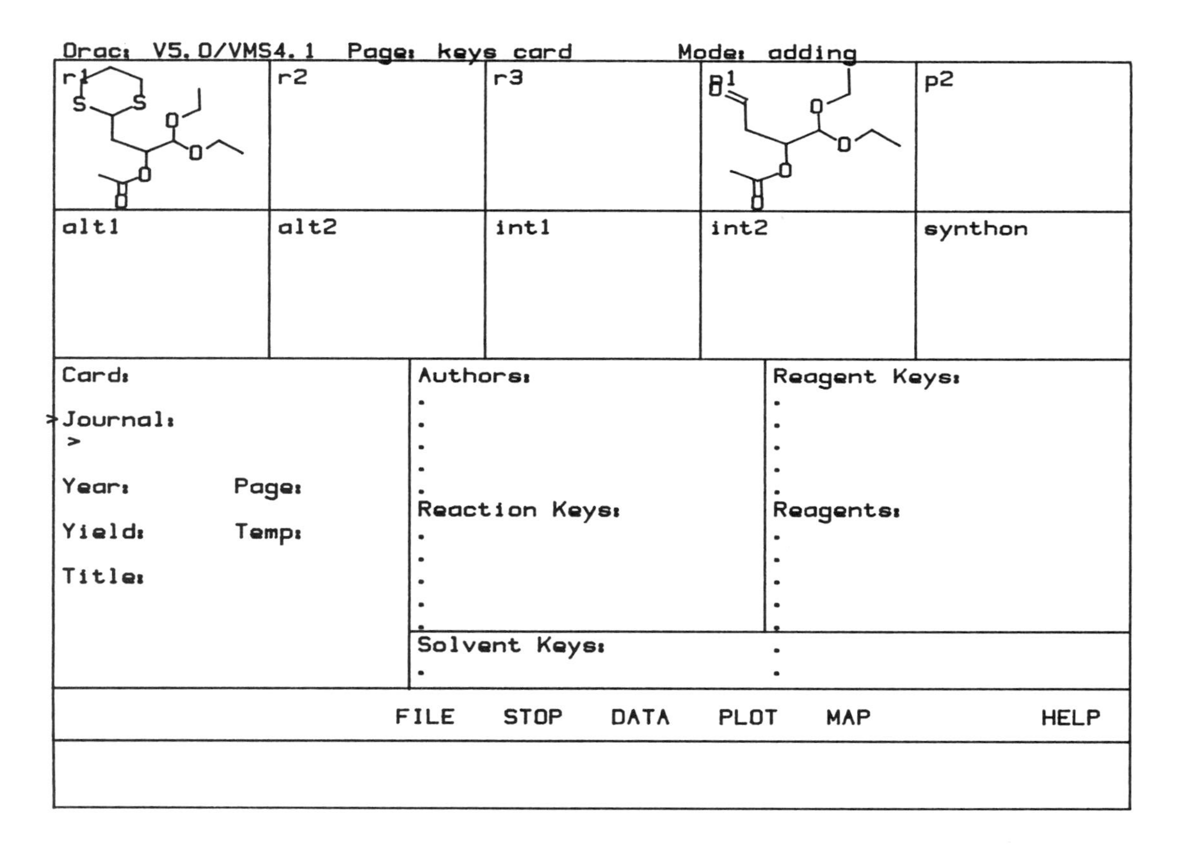
Orac: V5.0/VMS4.1 Page: keys card Mode: adding
r1
r2
r3
p1
p2
alt1
alt2
int1
int2
synthon
Card:
Journal:
Year:
Page:
Yield:
Temp:
Title:
Authors:
Reaction Keys:
Reagent Keys:
Reagents:
Solvent Keys:
FILE STOP DATA PLOT MAP HELP

Figure 4.

```
Orac: V5.0/VMS4.1  Page: keys library     Mode: searching
                  REACTION key-word library

PHOSPHONIUM
PHOSPHONYL CHLORIDE
PHOSPHORYLATION
PHOTOCHEMISTRY        Standard word: PHOTOLYSIS
                      Synonyms:       IRRADIATION
PHOTOLYSIS            Synonyms:       IRRADIATION, PHOTOCHEMISTRY
PHTHALIMIDE
PI BONDED COMPLEX     Standard word: PI COMPLEX
                      Synonyms:      PI COMPLEXATION, PI-COMPLEX
PI COMPLEX            Synonyms:      PI BONDED COMPLEX, PI COMPLEXATION,
                                     PI-COMPLEX
PI COMPLEXATION       Standard word: PI COMPLEX
                      Synonyms:      PI BONDED COMPLEX, PI-COMPLEX
PI-ALLYL
PI-COMPLEX            Standard word: PI COMPLEX
                      Synonyms:      PI BONDED COMPLEX, PI COMPLEXATION
PINACOL
PIPERIDINE

  SEARCH                         STOP              HELP

  Enter library index: PHOSPHONIUM
  Hit Space Bar for next page, or Return to stop
```

Figure 5.

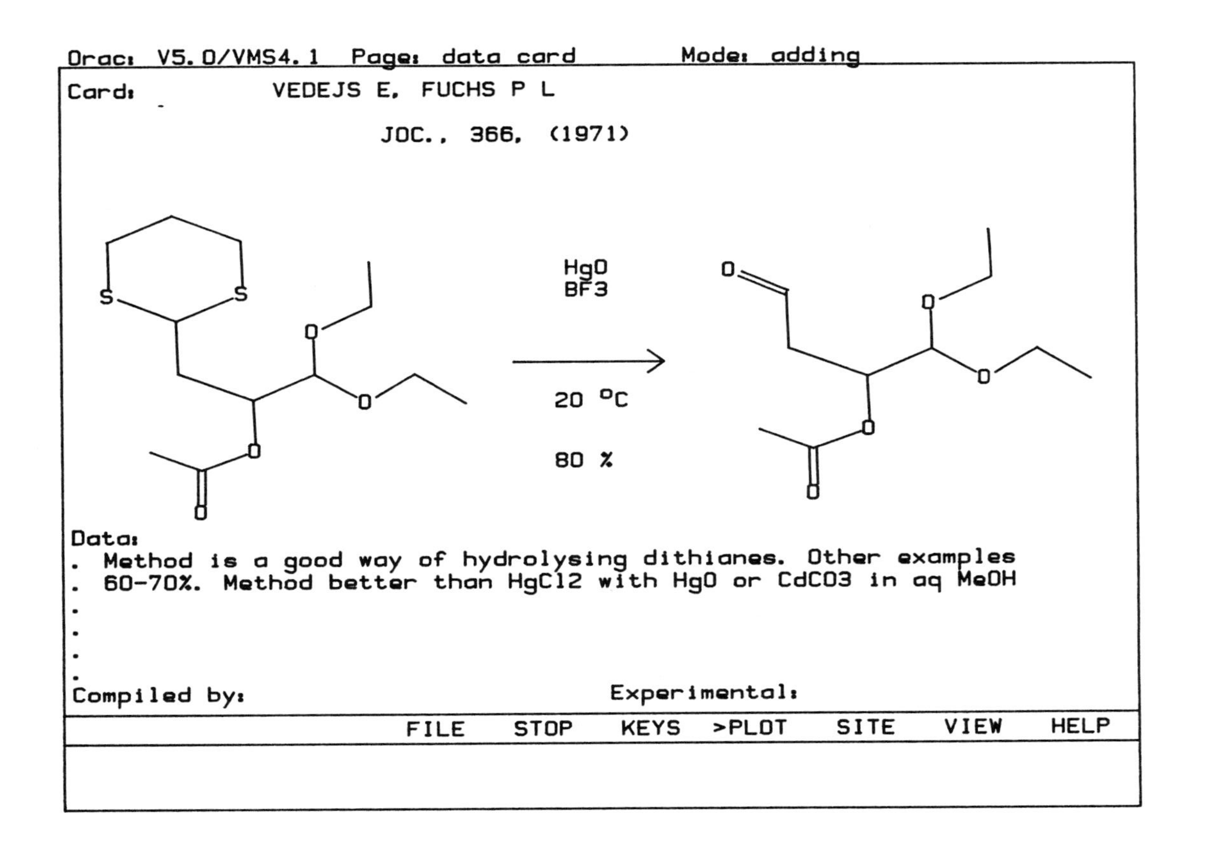
Orac: V5.0/VMS4.1 Page: data card Mode: adding
Card: VEDEJS E, FUCHS P L
JOC., 366, (1971)
HgO
BF3
20 °C
80 %
Data:
Method is a good way of hydrolysing dithianes. Other examples
60-70%. Method better than HgCl2 with HgO or CdCO3 in aq MeOH
Compiled by:
Experimental:
FILE STOP KEYS >PLOT SITE VIEW HELP

Figure 6.

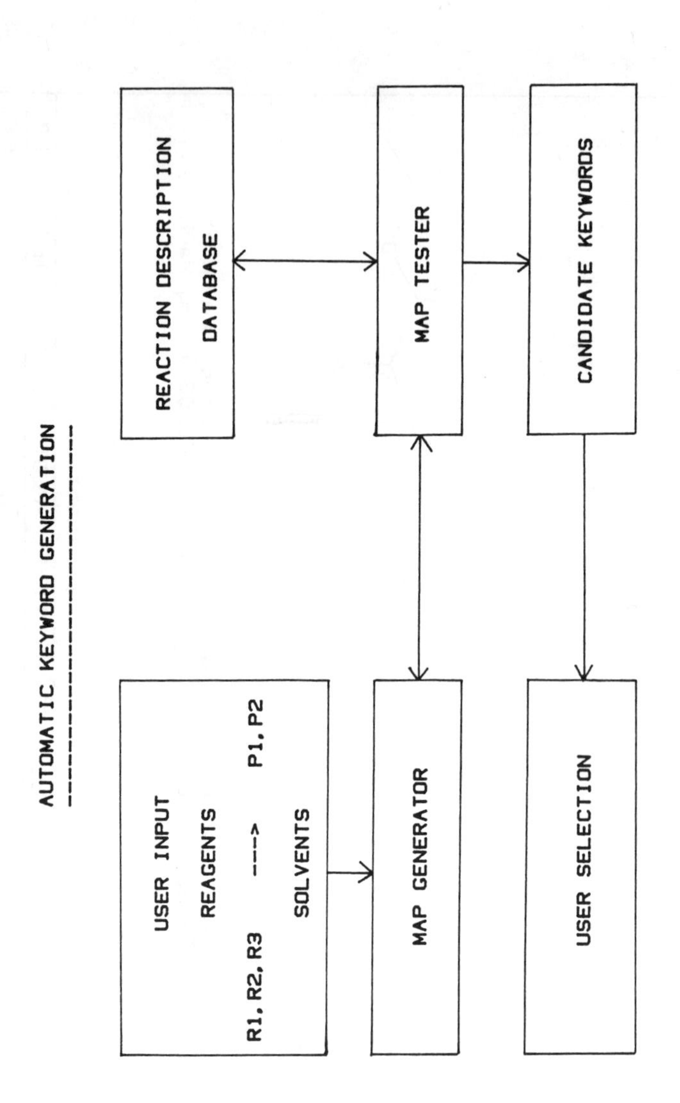
AUTOMATIC KEYWORD GENERATION
REACTION DESCRIPTION
DATABASE
MAP TESTER
CANDIDATE KEYWORDS
USER INPUT
REAGENTS
R1, R2, R3 ---> P1, P2
SOLVENTS
MAP GENERATOR
USER SELECTION

Figure 7.

Figure 8.

Generic class | Key substructure | Identified keywords

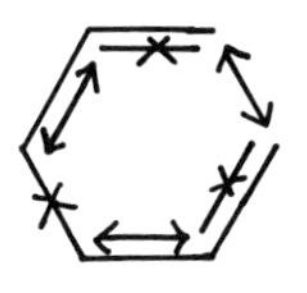

(3,3)sigmatropic

Specific class

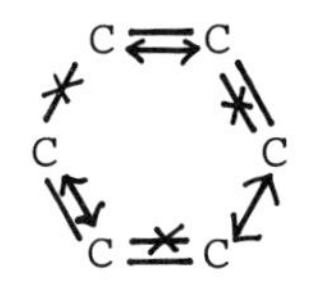

Cope rearrangement

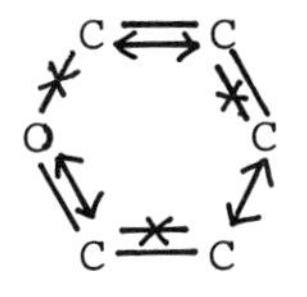

Claisen rearrangement

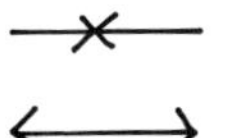

denotes a broken bond.

denotes a made bond.

15 Do we still need a classification of reactions?

G. Vladutz
Institute for Scientific Information

1. INTRODUCTION

The special symbols widely used in this paper for bonds made and broken by a reaction are correspondingly '⟷' and '—/—'. These symbols have been used previously in this author's 1974 book (Vladutz & Geyvandov, 1974) and in the 1977 report (Vladutz, 1977), both of which, for different reasons, have not been widely accessible. Using these symbols for a newly made or a broken single link the following symbols are derived for the formation (or breaking) of one or both links of a double or triple bond: '⟷̲', '⇔', ('=/=', '≠'); similarly for the formation or breaking of one, two or all three links of a triple bond the symbols: '≡⟷', '⇔̲', '⇔⟷' ('≡/≡', '≢', '≢'), will be used. It will be assumed that these symbols are available for graphical input and display on the terminals used for computerised retrieval of reaction information (CRRI).

We will try to answer the question posed in the title of this paper by considering the role of the classification of organic reactions for CRRI purposes. We will have in mind specifically the task of making possible the query input in the most unrestricted form, closest possible to the casual, offhand, off the cuff ways organic chemists are used to represent synthetic reactions. Some examples of this type of query presentation are given in Figure 1.

The casual reaction equations (Q1a) - (Q1d), which differ among themselves by different degrees of incompleteness describe in essence a same type of synthetic procedure traditionally designated as 'hydrocyanation of conjugated carbonyl compounds' (Organic Reactions, volume 25, 1977) which may be described most explicitly by the equation (Q1e) or most lapidarily, without using the reaction arrow, by (Q1f). The problem is how to recongise the practical equivalence of the so differently described classes of reactions

R1, R2 >C=C–C=O ⟶ R1, R2 >C(–C≡N)–CH–C=O (Q1a)

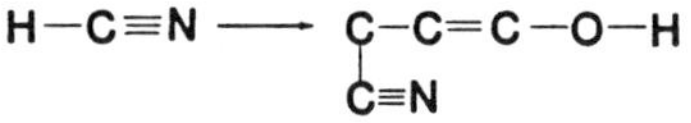

H–C≡N ⟶ C(–C≡N)–C=C–O–H (Q1b)

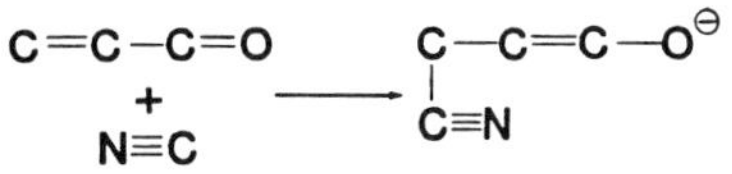

C=C–C=O + N≡C ⟶ C(–C≡N)–C=C–O$^{\ominus}$ (Q1c)

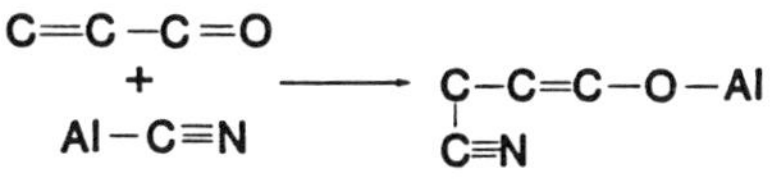

C=C–C=O + Al–C≡N ⟶ C(–C≡N)–C=C–O–Al (Q1d)

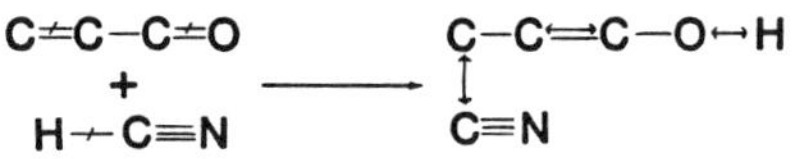

C≠C–C≠O + H↛C≡N ⟶ C(↕C≡N)–C⇌C–O↔H (Q1e)

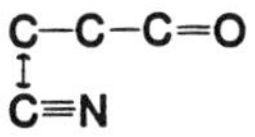

C(↕C≡N)–C–C=O (Q1f)

C(–O–H)(–Cl) ⟶ C=O (Q2a)

C(–O$^{\ominus}$)(–Cl) ⟶ C=O (Q2b)

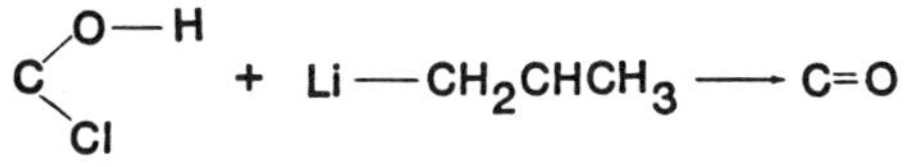

C(–O–H)(–Cl) + Li–CH_2CHCH_3 ⟶ C=O (Q2c)

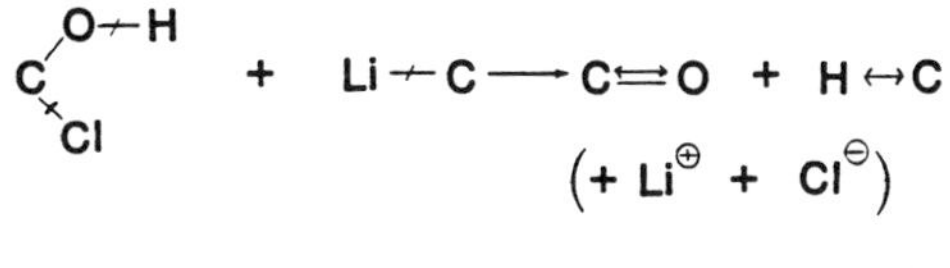

C(–O↛H)(↛Cl) + Li↛C ⟶ C⇌O + H↔C (Q2d)

(+ Li$^{\oplus}$ + Cl$^{\ominus}$)

Figure 1. Different forms of presentation of queries about reactions of hydrocyanation of conjugated carbonyl compounds (Q1a) - (Q1f) and reactions of ketone formation from α-chlorine-alcohol(ate)s (Q2a) - (Q2d) as examples of casual forms of query input.

and to retrieve this class of reactions in response to any of such queries. This becomes even more difficult when the data concerning the specific reactions belonging to the relevant class of reactions have been input into the system with similarly different degrees of incompleteness, which can be expected to be the case in the framework of any massive data gathering effort. Apparently the capabilities to solve such problems are not yet realised in the existing CRRI systems. Therefore it is the intention of this paper to examine some approaches to achieving such capabilities and thus to facilitate the course of the hoped for reaction CRRI ⟶ ICRRI, where ICRRI stands for <u>intelligent</u> CRRI.

As already noticed above any information query is equivalent to a definition (presented in a form understandable by the retrieval system) of the relevant to the query class of objects (records of documents or facts, in particular organic chemical reactions). Before the advent of the computerised information retrieval systems (CIRS) the only class characterisation tools usable for retrieval purposes were the traditional hierarchical classification headings according to which the corresponding document or fact files were structures. As an example, using the printed volumes of Beilstein only such classes of compounds can be retrieved in a direct way which correspond more or less precisely to some of the classes used in the Beilstein's hierarchical classification which uses the lack or the nature of cyclic systems for its main divisions of the universe of compounds, followed by the nature and number of functional groups used for the secondary and further subdivisions. The modern (IRS) information retrieval systems provide the capability to match much more sophisticated descriptions of objects (ranging for organic compounds from sets of specific fragments used as binary descriptors to the full record of structures) with the correspondingly formulated descriptions of requested classes of objects (classes of compounds defined by fixed structural fragment descriptors or arbitrary boolean combinations of arbitrary substructures). In such systems the hierarchical classifications are replaced by some kit of class definition tools which are used as query languages.

A desirable feature of any IRS is the ability to retrieve (with a reasonable degree of precision and recall) not only the objects exactly conforming to a given query but also, as second, third, etc. retrieved echelons the objects belonging to some 'natural' extensions of the class defined in the query. This gives the user the possibility to browse the portions of the database which are semantically in the more or less close vicinity of the exact class defined in the query. Thus not only the consequences of inexact query formulations, in particular of excessively restrictive formulations are corrected but help is provided for cases when the class defined by the query proves to be empty. Such natural extensions of queries are realised, in particular for the domain of compounds (Willett, 1982; 1983; Willett et al., 1985) using the 'best match' strategy, following which instead of retrieving only the objects belonging to a class defined by a given set of descriptors the classes most similar to the exact query class are also retrieved. The degree of similarity of classes and/or individual objects is defined on the basis of some similarity or dissimilarity function, the simplest of which is the degree of overlap between the corresponding sets of descriptors.

The nearest-neighbour searching makes it possible to completely avoid the usage of any special query language since queries can be formulated by presenting to the IRS the description of an individual object and requesting the output of objects which are most similar to this query object, the output being performed in the order of decreasing similarity to the query object. Instead of formulating the query as a single element class one can inquire about a class of objects by presenting the class by the simple enumeration of (a reasonably moderate number of) individual objects, thus also completely avoiding the usage of any special tools for class definition. (This is a 'give-me-more-like-these' type of search.) Talking about reactions this means that we would like to present to the IRS a record of individual reactions such as (Q1g) or a list of records of specific reactions such as the set [(Q1g), (Q1h), (Q1i)]

$$C_6H_5\underset{\underset{O}{\|}}{C}CH{=}CHC_6H_5 \longrightarrow C_6H_5\underset{\underset{O}{\|}}{C}CH_2\underset{\underset{C\equiv N}{|}}{C}HC_6H_5 \qquad (Q1_g)$$

$$CH_2{=}CH\underset{\underset{O}{\|}}{C}OCH_3 \longrightarrow N{\equiv}CCH_2CH_2\underset{\underset{O}{\|}}{C}OCH_3 \qquad (Q1h)$$

$$CH_3OCH_2CH{=}CHC{\equiv}N + H{-}C{\equiv}N \longrightarrow CH_3OCH_2\underset{\underset{C\equiv N}{|}}{C}HCH_2C{\equiv}N \qquad (Q1i)$$

and get as output the list of some several hundred conjugated hydrocyanation reactions as given, e.g., in the tables concluding the corresponding article in volume 25 of 'Organic Reactions'.

A common solution to both the above problem of query formulation for CRRI through the presentation of individual reactions or groups of such reactions as well as to the earlier mentioned problem of query formulation by using unrestrictively the most common, lapidary and most probably incomplete records describing classes of reactions (such as (Q1a), (Q2c)) is to derive some similarity measure equally applicable to individual reactions as well as to classes of reactions defined by such unrestricted means.

Another possible solution is the usage of some pre-established hierarchical classification system for reactions. Using such a system the first step has to be the determination of the most narrow subdivision which contains the specific reaction(s) presented in the query or the group of reactions described by the query's casual reaction scheme(s). This will enable the system to list as hits the full list of reactions belonging to the given subdivision of the reaction classification system. In order to obtain further echelons of similar reactions the higher level subdivisions of the classification system, containing the original subdivision are used consecutively.

We will examine both sketched above possible solutions starting with the latter one in Section 3. In this connection in Section 2 we will briefly review the existing hierarchical classification systems for organic reactions. In Section 4 an approach to the establishment of a similarity measure for specific reactions and unrestrictively defined classes of reactions will be suggested.

2. A BRIEF REVIEW OF MODERN HIERARCHIC STRUCTURAL CLASSIFICATION SYSTEMS FOR ORGANIC REACTIONS.

The concept of chemical reaction is by definition a more complex one than the concept of chemical compound. The characterisation of an individual instance of an organic reaction can include besides the full description of the ensemble of reactant and product molecules, many other elements related to the conditions under which the given reaction takes place and/or to its practical outcome, in particular the main product's yield and references to the co-occurring parallel reactions. When one assumes that reactions performed with the same ensemble of reactants under different conditions (solvents, temperature, pressure, etc.) but yielding the same ensemble of products are mere instances of the same individual reaction one takes a first step towards the classification of reactions on a purely structural basis. As a basis for the further grouping of individual reactions into classes, organic chemists have always used some characteristics of the structural changes resulting from the reaction, traditionally the nature of functional groups altered by the reaction and some general characterisation of the overall changes in the main participating molecules (e.g., 'substitution', 'addition', 'elimination', etc.), and later the nature of altered bonds. In Vladutz (1961; 1962/63) the more or less fuzzy concept of 'skeleton schemes' of reactions, corresponding to succinct structural equations of the type of (Q1a) - (Q1g) was formalised as a pair of substructures correspondingly of the reagent and product molecules, consisting of all the 'key' atoms having some bonds altered in the course of a given reaction; a 1-1 correspondence is established between the atoms of these substructures. This pair of substructures describes such a substitution operation of the first subgraph of the pair by the second subgraph of the pair which being performed on the graph consisting of reactant molecules will result in the transformation of that graph into the graph consisting of product molecules (Vladutz, 1961). This formalised concept of the skeleton scheme of a reaction was further described and illustrated in a slightly simplified form in Vladutz (1963) and Vladutz & Finn (1963) along with practical ways for the approximate linear encoding of skeleton schemes and the possibilities for the usage of a library of such schemes for computerised synthesis design. Kiho (1972) suggested a more compact way for the representation of individual reactions, as well as of reaction skeleton schemes (RSS), which makes the usage of the correspondence tables between the atom numbers of reactant and product molecules (or between the atom numbers of the left and right hand sides of the RSS) unnecessary by representing the (key) atoms of both reactant and product molecules only once (instead of repeating them in the left and right hand sides of the reaction equation or of the RSS). The so obtained connected multigraph with labelled vertices (atoms) and edges (bonds) makes use of different edge labels for bonds which are either not affected or are altered by the reaction. We will use the special symbols introduced in the first section as labels for the altered bonds. Such compact representations for individual reactions (as well as for their RSS, see below), using the one-time display of the ensemble of atoms making up the molecular species involved in the reaction were widely used in the Vladutz (1974) book and were designated later (Vladutz, 1977) as 'superimposed reaction graphs' (SRG). By omitting from a SRG the vertices which do not participate in any altered bond a subgraph called 'superimposed reaction skeleton graph' (SRSG), consisting of all the key atoms is obtained, which is a most compact representation of an RSS.

The SRG of the individual reaction (Q1i') and the corresponding SRSG (Q'1i) are given below as an example:

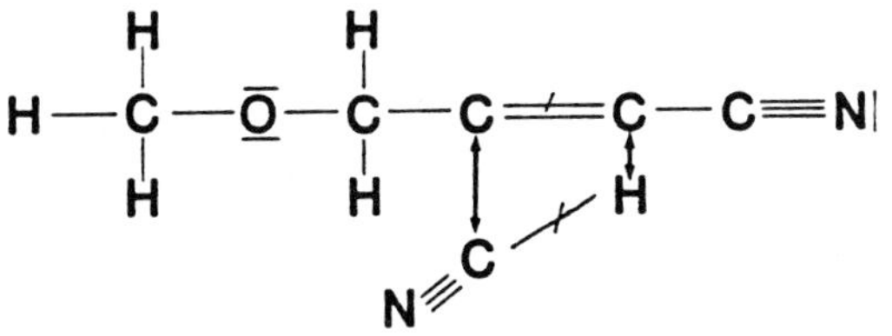

(Q1i')

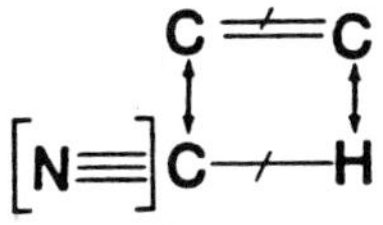

(Q'1i)

Since the H-C≡N molecule can be represented alternatively as an ion pair there are correspondingly some completely legitimate alternative forms of these expressions:

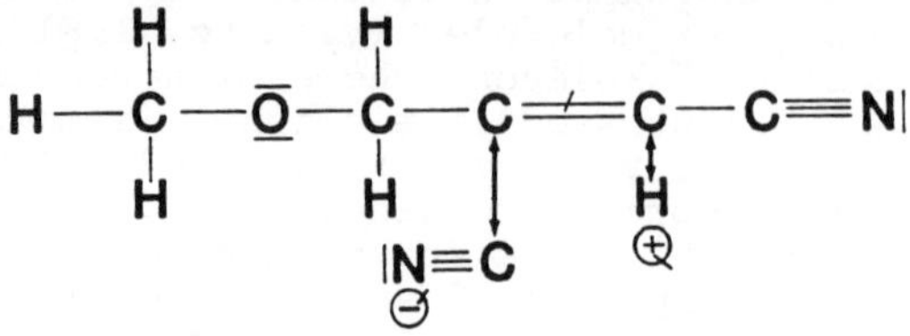

$$(Q1i'_1)$$

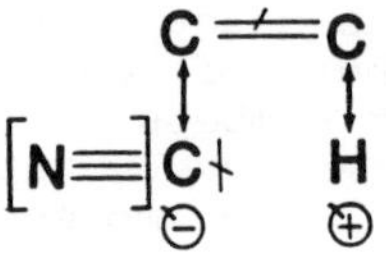

$$(Q'1i_1)$$

Besides illustrating some of the problems which can be involved in the uniqueness of SRGs and SRSGs this example introduces some more new special symbols, in particular the symbol '—/—' placed on an atom (representing a loop of the double labelled multigraph) designating such an unshared electron pair on the corresponding atom of a reactant molecule which is no longer present on that atom after the reaction. The corresponding symbol '⟷' placed on an atom for designating an unshared electron pair which appeared on it in the product molecule is used in the SRSG for (Q1c):

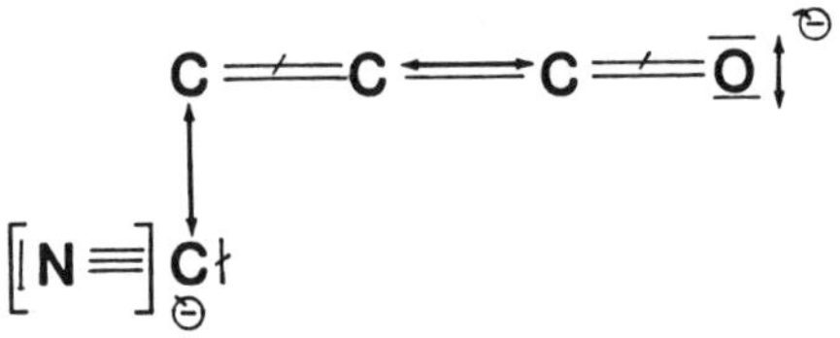

$$(Q'1c)$$

In the above two examples the usage of symbols (also representing loops consisting of labelled edges) ⊕, ⊖, ⊕, ⊖ for positive and negative charges present correspondingly only in reagent or product molecules is also illustrated; it is assumed that all the molecules participating as reagents or products of synthetic reactions are stable species with all the atoms having stable electron configurations. In these examples the N atom of the cyano-group is not a key atom for these reactions and therefore they do not belong legitimately to (Q'1i) or (Q'1c); they are displayed in square brackets for easier visualisation.

The above concept of SRSG is in many respects equivalent to the concept of reaction matrices (R-matrices) developed independently in the framework of what has become known as the DUGUNDJI-UGI model of constitutional chemistry (Dugundji & Ugi, 1973; Ugi et al., 1979a; Brandt et al. 1981, 1983, 1984; Bauer & Ugi, 1982; Brandt & von Scholley, 1983; and literature cited therein). More precisely the so called irreducible R-matrices, R^I (Brandt et al., 1981) are equivalent to a SRSG in which the nature of the key atoms is not specified and no unaltered bonds are represented (in other words —/—, =/=, ≡/≡ are represented as —/—, and ⟷, ⟺, ⟺ are represented as ⟷). The actual SRSGs are closely equivalent to Brandt's 'reaction cores', expressed as 3-tuples (R^C, a^C, B^I), consisting of a canonical irreducible R-matrix, R^C, a vector, a^C, specifying the nature of the key atoms and a matrix, B^I, indicating those bonds and unshared electrons that are not changed by the reaction. The slight difference is that, unlike the reaction cores, the SRSGs do not include the unchanged bonds existing between key atoms.

In Brandt et al. (1981) a hierarchical classification of reactions was outlined, the most general classes of which correspond to groups of reactions with identical R^C. Further subdivisions of such classes are created using consecutively the values of a^C and B^I and then by considering levels of ascending neighbour spheres of key atoms and thus taking consecutively in consideration (for all key atoms simultaneously) the descriptors of these first, second and further levels of neighbouring atoms.

The similar in many respects classification of reactions based on SRSGs outlined in Vladutz (1963) and further developed in Vladutz & Geyvandov (1974, Section 14.4) takes as primary divisions the groups of reactions designated as SRSG-classes, having the same (i.e., isomorphic) SRSGs. For the creation of further subdivisions the SRSGs are gradually augmented along two different lines of detailing. The first of these detailing procedures generates consecutively augmented SRSGs denotated as $SRSG_1$, $SRSG_2$, $SRSG_3$, etc., called SRSGs of different levels, by making more and more specific the lack or the presence and the nature of the unchanged link symbol, denotated by 'Ω', (i.e., an additional, specially labelled edge of the multigraph) is inserted between all such key atom pairs the shortest path between which does not include any third key atom. In $SRSG_2$ the Ω-links are replaced either by the specific bond(s) existing between the key atoms if that happens to be the case or by the symbol 'ω', denotating any chain of atoms, containing not less than one atom; in $SRSG_3$ the ω-links are replaced by the specific shortest chain(s) of atoms linking the given pair of key atoms. The bonds between these atoms are also included. Figure 2 gives some examples of SRSGs of different levels.

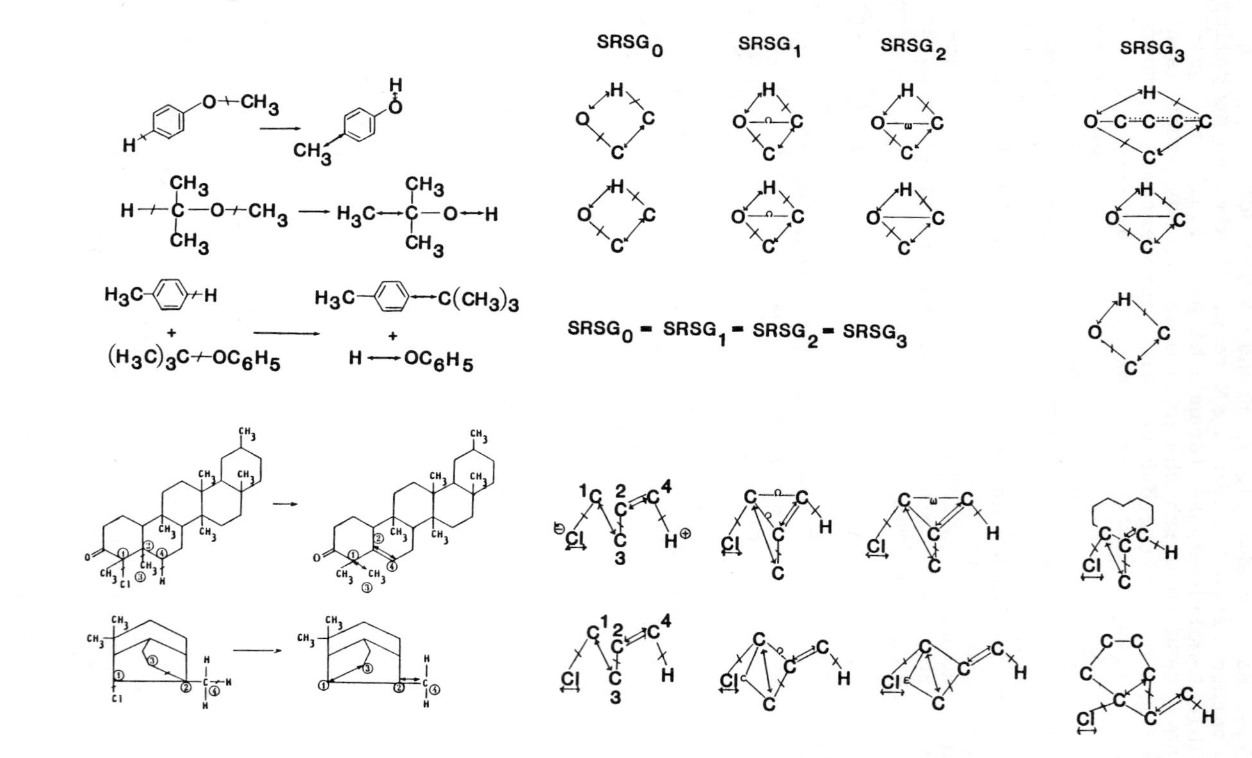

Figure 2. Examples of SRSGs of different level. Shown are two groups of reactions which have correspondingly the same zero level SRSG$_0$ but different higher level SRSGs.

A different second type of detailing procedure augments a reaction's $SRSG_{\underline{i}}$ (where $\underline{i}$ can be any value from 0 to 3) by consecutively including the non-key atoms located in the immediate and then also in the more remote vicinity of key atoms. This detailing procedure generates the so-called reactions type schemes (RTS) and, correspondingly, their compact representations in the form of superimposed reaction type graphs (SRTG) of different degrees:

$$SRTG^{o}_{\underline{i}},\ SRTG^{1}_{\underline{i}},\ SRTG^{2}_{\underline{i}},\ \text{etc.}$$

where

$$SRTG^{o}_{\underline{i}} \equiv SRSG_{i}.$$

Without going into all the specifics it will be enough to consider here that the SRTGs of different degrees differ by the inclusion of the non-key atoms located correspondingly at distance 1 (immediately connected), 2, 3 etc. from the key atoms. Examples of SRTGs of different degrees are shown in Figure 3.

An open question for this type of hierarchical structural classification is the exact number and the specific list of their main divisions. In the case of the SRSG-based classification this boils down to the number and specific list of valid (i.e., well formed) SRSGs; in the case of Brandt's R^{C}-matrices-based classification we are talking about the variety of legitimate (well-formed) R^{C}-matrices. A procedure for the algorithmic generation of well formed SRSGs was suggested earlier (Vladutz, 1977) for the particular but frequent case of cyclic SRSGs corresponding to quasi-cyclic electron shift patterns. In Figure 4 a more exhaustive reaction skeleton scheme generation procedure is briefly outlined and its results shown for the case of the relatively few SRSGs involving less than 4 key atoms taken from the set (C, H, N, O, Cl). This algorithmic procedure starts by generating all the connected graphs consisting of less than $\underline{k}$ atoms; for $\underline{k} < 4$ the graphs are represented in the first column of Figure 4; the edges (including possible loops) of each generated graph are then labelled with the symbols of altered bonds in all those possible ways which preserve the electronic balance, do not exceed the valency limits of the selected species of key atoms, correspond to stable for the given atomic species electronic configurations and do not exceed for each participating molecular species the admissible number of charged atoms (taken here as = 2) with charges in the admissible range (+1, -1). A program built along these lines can be easily written for generating an exhaustive list of $SRSG_{o}$-s. For $\underline{k} < 8$, which according to Bart & Garagnani (1977) is sufficient for practical purposes, one can guess that the number of well formed $SRSG_{o}$-s will not exceed a few thousands, with probably fewer number of the corresponding, more general in nature R^{C} matrices in the 200-300 range. (A significant percentage of these generated well formed reaction schemes would not be acceptable due to stereochemical limitations.)

In virtue of the above considerations we will assume that reasonably full listings of the main divisions of both above mentioned hierarchical classification systems for reactions can easily be made accessible.

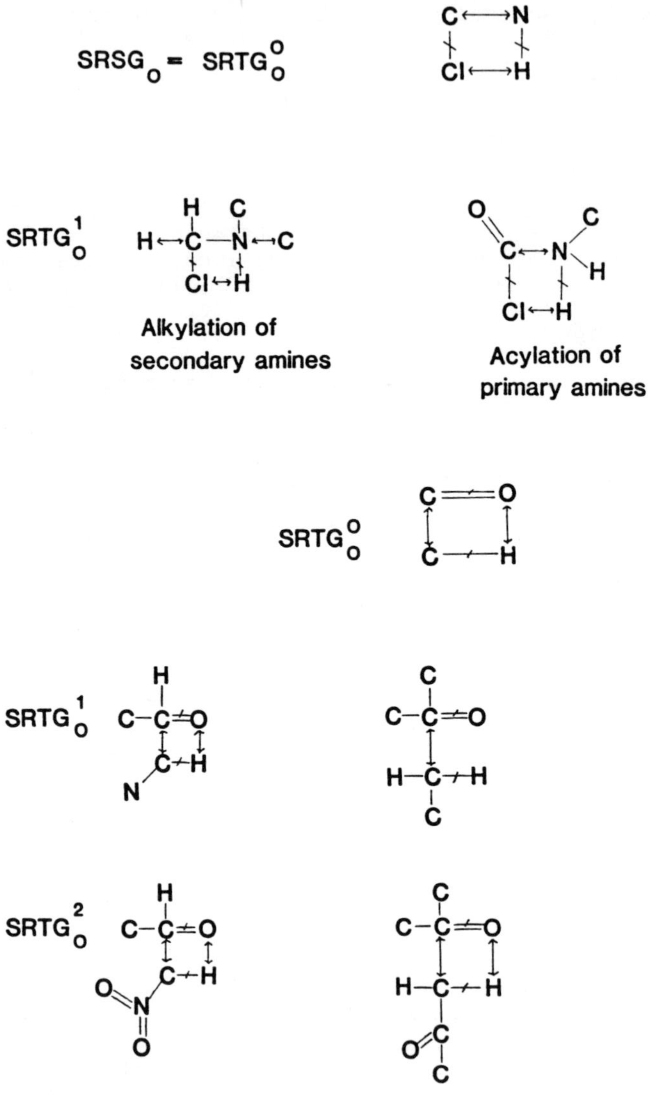

$SRSG_O = SRTG_O^O$

$SRTG_O^1$

Alkylation of secondary amines

Acylation of primary amines

$SRTG_O^O$

$SRTG_O^1$

$SRTG_O^2$

Figure 3. Examples of SRTGs of different degree. In the first example the $SRTG_O^1$s differentiates between an aldehyde or ketone participant of the condensation reaction and $SRTG_O^2$ reveals the nature of the α-functional group activating the C-H bond.

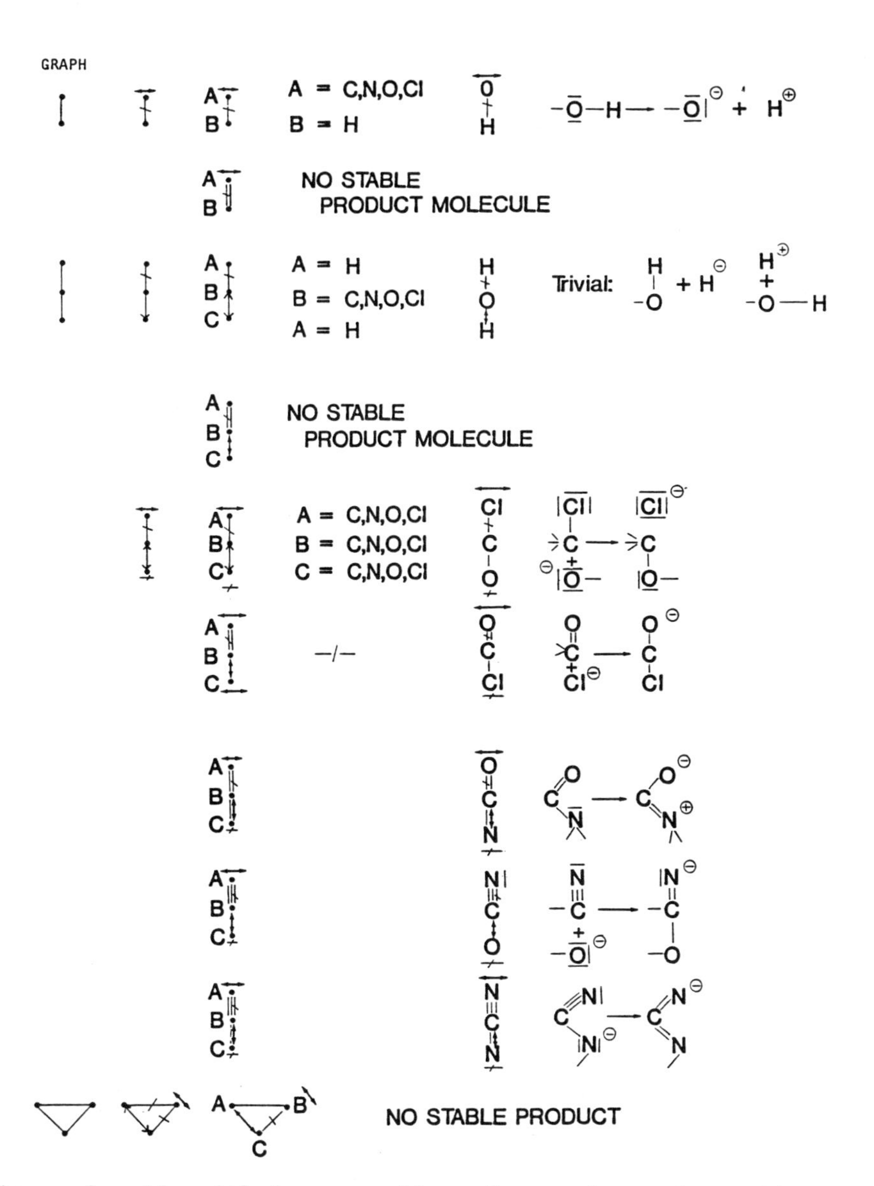

Figure 4. Algorithmic generation of superimposed reaction skeleton graphs (SRSG) illustrated for the case of less than 4 vertices (atoms from the set: C, H, N, O, Cl). The first column enumerates the graphs, the second column gives the electronically balanced schemes consisting only of altered bonds, in the fourth column possible unchanged bonds are added and finally an example of the so produced generalised reaction scheme is given.

3. THE APPLICABILITY FOR ICRRI OF HIERARCHICAL CLASSIFICATIONS OF REACTIONS BASED ON STRUCTURAL CHANGES.

As outlined at the end of the introduction the easily conceivable general way for making use of hierarchical classifications for intelligent retrieval is to translate each query into the description of the most narrow subdivision of the classification which still satisfies the query's requirements and then output as hits the objects assigned to this subdivision, followed by the objects classified in the immediate superclass and so on.

In the specific case of reactions, such an approach involves the translation of succinct or casual queries as well as of queries formulated as lists of specific reactions into the descriptions of the corresponding reaction classes, i.e., into SRSGs or R^C matrices. A common feature of both above types of class descriptions is that they are based on stoichiometrically completely balanced reaction equations. However, only few if any of the real life queries presented to a CRRI system can be expected to fulfill this condition. More than that, in the real life conditions of the creation of any large enough reaction database data needed to compile fully balanced equations will be available only for a small fraction of specific recorded synthetic reactions. Therefore the best one can do on the basis of the available data is to generate some, often partially incomplete expressions of the reaction class descriptions and then to best match these partial descriptions with the computer generated sets of well formed full class descriptions. More precisely we are talking about the translation of the more or less arbitrarily formulated reaction retrieval queries on the basis of the maximum information contained in them into partial expressions as closely as possible related to the $SRSG_o$-s of relevant specific reactions. For the above query examples the results of such translations may be the expressions shown in Figure 5.

The superimposed schemes shown in the right column of this figure are obtained from the original query skeleton schemes by an algorithmic procedure which involves mainly the detection of the maximal common subgraphs (MCS) shared by the two sides of the skeleton scheme, followed by the superimposition of the two structures according to their MCS. The MCS is shown in a solid box, whereas the parts of the superimposed graph which are outside the MCS are shown in dotted or dashed boxes depending upon their occurrence in the left or right hand side of the original scheme respectively. For some existing procedures of MCS detection see McGregor & Willett (1981), Wong & Akinniyi (1983), Brandt et al. (1984), Brandt & Wochner (1984) and, for procedures approximating MCS detection see Lynch & Willett (1978), Jochum et al. (1980).

If ICRRI should be based on the usage of a hierarchical classification then following the automatic translation of the queries into superimposed graph expressions the next step is to determine the lowest level classification sub-

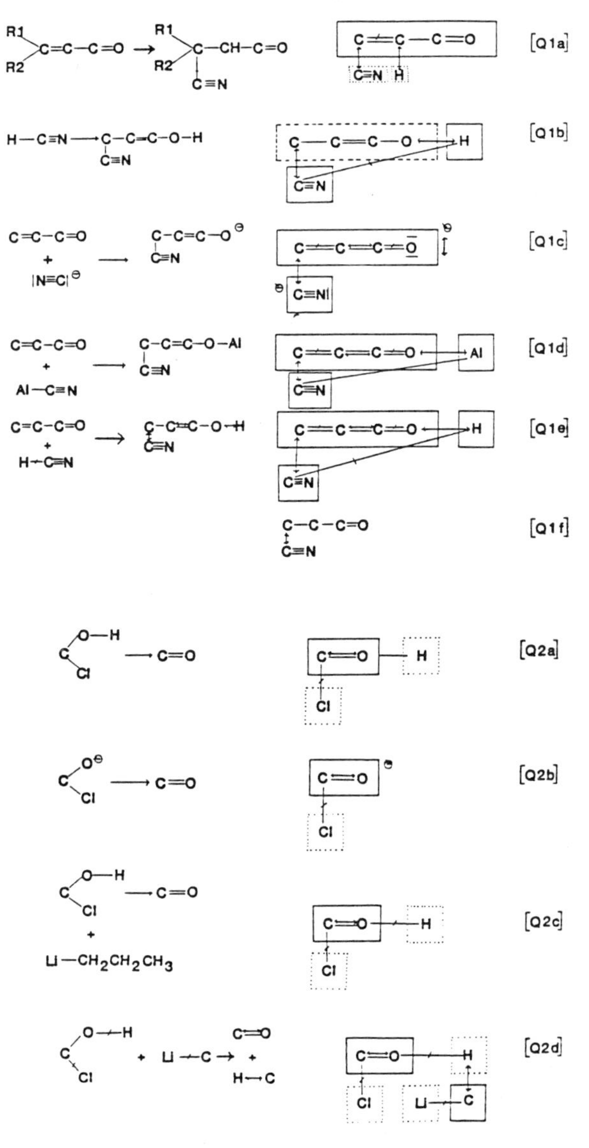

Figure 5. Algorithmically generated generalised super-imposed reaction graphs (GSRGs) obtained from casual reaction queries (Q1a) - (Q2d) on the basis of the detection of the maximum common sub-graphs (MCS), which are shown in solid boxes. The GSRG portions outside the MCS are shown in dotted or dashed boxes according to their occurrence in the left or right hand side of the original reaction schemes.

division in which a group of reactions characterised by some given superimposed structural scheme is entirely contained. Such a determination has to be performed by some algorithmic comparison between such a structural scheme and the automatically generated structural schemes describing the classification's subdivisions. Such a comparison, resulting in the automatic assignment of a query to a specific subdivision of the classification can be, perhaps, best performed on the basis of some similarity or dissimilarity function.

In order to perform the final stages of the ICRRI process involving the output of specific reactions assigned to the given classification subdivision and then consecutively of those assigned to the higher level broader subdivisions containing the originally determined most narrow class one has to assume that all the reactions of the database have been (automatically) assigned to the respective classification subdivisions.

It is hard to believe that for any large enough reaction database sufficient data (in particular stoichiometrically fully balanced equations) can be compiled which would make it possible, even using some flawlessly performing reaction site detection procedure, to automatically derive for each specific reaction of the database the precise expressions determining the assignment of each reaction to the corresponding classification subdivision. Rather more realistic is to presume that such assignments may be performed as best matches against some pre-existing formalised class descriptions using again some similarity or dissimilarity function. Thus the essence of the ICRRI using a hierarchic classification of reactions based on structural changes would ultimately boil down to the determination of the degree of similarity (or dissimilarity) between queries and specific reactions. The role of the proper classification would be to discretely quantify the discernable degrees of similarity.

4. ICRRI BASED ON THE NEAREST NEIGHBOUR CONCEPT.

As noted above in the Introduction the alternative to the usage of some hierarchical classification is the direct application of the best match strategy using some powerful enough similarity measure, equally applicable to specific reactions as well as to classes of reactions defined in the various most common and casual way. Since, as we just argued, the classification based approach cannot do without best matching procedures one is led to the conclusion that this second alternative is the only viable one.

The above outline procedure illustrated in Figure 5 and based on the detection of the MCS of two arbitrary graphs is equally applicable to the most succinct forms of presentation of queries expressed as casual structural reaction schemes as well as to the most rigorous fully stoichiometrically balanced specific reaction equations. In the latter case it results in the precise expression of some $SRSG_k$ or $SRTG_k^1$, as defined above. In the former cases or in the absence of full reaction equations the procedure delivers something we propose to call 'generalised superimposed reaction graph' (GSRG) representing a partial approximation to some $SRTG_k^1$.

Therefore it remains to choose a similarity function defined on arbitrary pairs of GSRGs which would perform in accordance to the chemist's intuition concerning the degrees of similarity and/or dissimilarity of organic reaction. An approach to the construction of such a similarity measure is outlined below.

Let $V(G_k)$ be a function which assigns to any GSRG = G_k or any subgraph G_k of some GSRG a numeric value and let $(G_i \cap G_j)$ denote the MCS of GSRGs G_i and G_j.

The suggested form of the sought similarity measure S ($0 \leqslant S \leqslant 1$), applicable to arbitrary pairs G_i, G_j of GSRGs is:

$$S(G_i, G_j) = \frac{V(\langle G_i \cap G_j \rangle)}{\min(V(G_i), V(G_j))}$$

This is a generalisation of the similarity measure called Overlap coefficient

$$\frac{|X \cap Y|}{\min(|X|,|Y|)}$$

used in particular in information retrieval (see van Rijsbergen, 1979), where X and Y are sets of document or object descriptors.

Without going here into details the value function V, indicating the importance from the point of view of the overall similarity of some identical or non-identical portions of two GSRGs, can be briefly characterised as consisting of the product of some partial values assigned to bonded atom pairs with the highest partial values for the altered bonds and essentially less and gradually decreasing values for unchanged bonded pairs of atoms located at increasing distances from the key atoms (reaction site). The value of V for an empty graph is zero; therefore for both cases when either G_i is a subgraph of G_j or G_j is a subgraph of G_i one obtains $S(G_i, G_j) = 1$. In those cases when some graph G_i is 'almost' a subgraph of some other graph G_j the similarity measure $S(G_i, G_j) = S(G_j, G_i)$ remains close to 1. Such is the case for all possible pairs of the above GSRGs (Q1a - Q1f) and (Q2a - Q2d).

The most practical overall strategy for reaction database organisation using the suggested similarity measure appears to involve the batch mode processing of the reaction data input file for determining for each specific input reaction a number $< k$ (k = 3 - 4) of other specific reactions of the file which are for the given reaction the most similar ones. The database is then created as a network in which neighbouring objects display a high degree of similarity. A similar approach has been recently suggested (Vladutz, 1984) to the creation of a large document database using interdocument similarity measures based on bibliographic coupling.

For any query the closest (according to the proposed similarity measure) neighbour is found in this network and its vicinity (first immediate then more and more remote) is then gradually explored.

The usage of the proposed similarity measure as the basis for some hierarchical clustering procedure applied to a large enough set of specific reactions should lead to the automatic re-creation of some hierarchical reaction classification system which should be close enough to the existing ones. However, the usage of such a reaction classification system would not be needed for searches. The nearest-neighbour browsing in the region of the network identified by the similarity with the query achieves the goals of best match searches which may come close enough to the results of intelligent retrieval as performed by humans.

CONCLUSION

The examination from the point of view of utility for ICRRI of the modern structural changes based hierarchical classifications of reactions led us to the conclusion that a more practical and at the same time theoretically more appealing approach to ICRRI, which makes full use of the ideas embodied in such classifications, may be based on the best match/nearest neighbour search strategy. An algorithmically derivable similarity measure, equally applicable to rigorous as well as more casual descriptions of specific reactions and more or less general reaction geared queries has been suggested for this purpose.

REFERENCES

Bart, C. J. and Garagnani, E. (1977). Organic Reaction Schemes and General Reaction-Matrix Types, II. Basic Types of Synthetic Transformations. *Zeitschrift fur Naturforschung, B. Anorganische Chemie*, *Organische Chemie*, 445-464.

Bauer, J. and Ugi, J. (1982). Chemical Reactions and Structures without Precedent, Generated by Computer Program. *J. Chem. Research* (*M*), 3101-3109.

Brandt, J. (1981). Ein mathematisch begründete Hierarchie der organischen chemischen Reaktionen und deren theoretische und praktische Anwendungen. Habilitationschrift, Technische Universität München, Anhang A.

Brandt, J., Bauer, J., Frank, R. M. and von Scholley, A. (1981). Classification of Reactions by Electron Shift Patterns. Chemica Scripta, vol. 18, 53-60.

Brandt, J. and von Scholley, A. (1983). An Efficient Algorithm for the computation of the Canonical Numbering of Reaction Matrices. Computers and Chemistry, 7(2), 51-59.

Brandt, J., von Scholley, A. and Wochner, M. (1984). Making Chemical Reaction Data Accessible to the Non-Chemist. Computer Physics Communications, 33, 197-203.

Brandt, J. and Wochner, M. (1984). Private Communication.

Dugundji, J. and Ugi, J. (1973). Algebraic Model of Constitutional Chemistry as a Basis for Chemical Computer Programs. Top. Curr. Chem., 39, 19-64.

Jochum, C., Gasteiger, J. and Ugi, J. (1980). The Principle of Minimum Chemical Distance (PMCD). Angewandte Chemie Int., 19, 495-505.

Kiho, Y. K. (1972). Formal'noe Opredelenie nekotorykh ponyatiy koichestvennoy organicheskoy khimii. (The Formal Definition of some concepts of Quantitative Organic Chemistry). Organic Reactivity (Tartu), 8(2).

Lynch, M. F. and Willett, P. (1978). The Automatic Detection of Chemical Reaction Sites. J. Chem. Inf. and Comp. Sci., 18, 154-159.

McGregor, J. J. and Willett, P. (1981). Use of a Maximal Common Subgraph Algorithm in the Automatic Identification of the Ostensible Bond Changes Occurring in Chemical Reactions. J. Chem. Inf. and Comp. Sci., 21, 137-140.

van Rijsbergen, C. J. (1979). Information Retrieval (Second Edition), Butterworths, London.

Ugi, J., Bauer, J., Brandt, J., Friedrich, J., Gasteiger, J., Jochum, C., Lemmen, P. and Schubert, W. (1978). Deductive Solution of Chemical Problems by Computer Programs on the Basis of a Mathematical Model of Chemistry. Pure and Applied Chemistry, 50, 1303-1318.

Ugi, J., Bauer, J., Brandt, J., Friedrich, J., Gasteiger, J., Jochum, C. and Schubert, W. (1979). New Applications of Computers in Chemistry. Angewandte Chemie, 18(2), 111-123.

Vladutz, G. (Vleduts, G. E.) (1961). O nekotorykh informatsionnologicheskikh zadachakh iz oblasti organicheskoy khimii. (Concerning some informational-logical problems from the domain of organic chemistry). Doklady na konferentsii po obrabotke informatsii, mashinnomu perevodu i avtomaticheskomy chteniyu teksta, 5, Moscow, 52 pages.

Vladutz, G. (1962). Ob odnoy sisteme klassifikatsii i kodirovaniya organicheskikh reaktsiy. Voprosy mekhanizatsii i avtomatizatsii informatsionnykh rabot. VINITI, Moscow. (in Russian; see below [Vladutz, 1963]).

Vladutz, G. (1963). Concerning one System of Classification and Codification of Organic Reactions. Information Storage and Retrieval, 1, 117-146.

Vladutz, G. (1977a). Compact Representation of Chemical Reactions for Automated Retrieval and Generation. Abstracts of Papers Presented at the 174th Meeting of the American Chemical Society, Chicago, Il., p.18.

Vladutz, G. (1977b). Development of a Combined WLN/CTR Multilevel Approach to the Algorithmic Analysis of Chemical Reactions in View of their Automatic Indexing. British Library R & D Dept.,Report 5399.

Vladutz, G. (1984). Bibliographic Coupling and Subject Relatedness, Proceedings of the 47th ASIS Annual Meeting, 21, 204-207.

Vladutz, G. (Vleduts, G. E.) and Finn, V. K. (1963). Creating a Machine Language for Organic Chemistry. Information Storage and Retrieval, 1, 101-116.

Vladutz, G. (Vleduts, G. E.) and Geyvandov, E. A. (1974). Avtomatizirovannye informatsionnye sistemy dlya khimii (Automated information systems for chemistry), Nauka, Moscow, 312 pages.

Willett, P. (1982). The Calculation of Intermolecular Similarity Coefficients Using an Inverted File Algorithm. Analytica Chimica Acta, 138, 339-342.

Willett, P. (1983). Some Heuristics for Nearest-Neighbour Searching in Chemical Structure Files. J. Chem. Inf. Comp. Sci., 23(1), 22-25.

Willett, P., Winterman, V. and Bawden, D. (1985). Implementation of Nearest Neighbour Searching in an Online Chemical Structure Search System. Submitted for publication.

Wochner, M. (1984). PhD thesis, Technische Universität München.

Wong, A. K. C. and Akinniyi, F. A. (1983). An Algorithm for the Largest Common Subgraph Isomorphism using the Implicit Net. 1983 Proceedings of the International Conference on Systems, Man and Cybernetics, 1, 197-201.

16 A recursive reaction generator

J. Brandt and K. Stadler
Technische Universität München

SUMMARY

The R-matrix formalism of the DUGUNDJI-UGI-model is a mathematically consistent and structurally complete representation of an individual chemical reaction. However, the solution of the 'reaction generator' paradigm would, mathematically, require the addition of each member of the complete set of permutations of an R-matrix to the given problem structure, as represented by a BE-matrix. This is not only technically unfeasible, but would also generate an unacceptable number of chemically meaningless solutions. Therefore, the permutations have to be generated step by step, so that unsuccessful branches can be terminated immediately as they occur. This strategy, hitherto, required implementation of individual pieces of program code for each individual R-matrix. Existing synthesis planning programs use reaction generators of this type, and are therefore limited to a small set of reaction patterns.

The R-string representation of R-matrices in connection with the canonicalisation rules that we originally developed for use in reaction documentation and chemical didactics, enables us to design a universal reaction generator module. Here, the code contains only the strategy, while the R-matrix is passed to it as a piece of data in the form of a canonical R-string. The R-string directs the execution of the strategy, but does not modify the code. The reaction generator is coded as a recursive PASCAL procedure. It checks for the validity of each completed step before scanning the atoms of the problem structure for candidates of the next step. The set of computed solutions is complete. In the present implementation, no provision is yet made for recognition of symmetry, or in-line elimination of duplicate solutions. The universal reaction generator gives the user complete freedom to evaluate any reaction pattern, or collection of reaction patterns, as long as these are valid by the rules of the DUGUNDJI-UGI-model. Further flexibility is provided for the free definition of chemical boundary conditions: reaction-invariant bonds, different sets of valence schemes for atoms that are members of the reaction core (thus allowing closed-shell-, radical- or complex-chemistry), and unspecified atoms in the reaction core.

1. INTRODUCTION

1.1 Reaction Operators

Anybody who starts thinking about the computer coding of chemical reactions will sooner or later perceive the numerical representation of a reaction as an operator which acts on an operand, or set of operands. These operands are numerical representations of chemical structures which are to be converted into other chemical structures. (For convenience, we will call these operands 'problem structures'.)

Consequently, one of the central pieces of a reaction modelling program or of a synthesis planning system is the reaction generator. This is a piece of code which executes the operation that transforms the computer code for the reactands into that of the products.

1.2 Reaction Descriptors

The theme of this conference is: Modern Approaches to Chemical Reaction Searching. 'Searching' implies a fairly different problem from that of the computer coding of reactions which is more of a descriptive nature and which aims at the problem of defining, storing and retrieving a descriptive representation of a reaction.

I would now like to invite you to follow me on a path that started with the operational aspect of a representation of reactions, and was at first directed towards deriving a descriptor from this representation. But then, when this first goal was achieved, we discovered that this descriptive representation yields a surprisingly efficient implementation of the operational aspect of a reaction. This path back to the operator was not a closed loop like a cycle, but resembled rather more a spiral in the sense that the operator now appears in the form of a parameter that can be executed by a computer code. By contrast, the original matrix representation was, although mathematically exact, not very suitable for direct conversion into a computer code. The exciting aspect of this new approach is that the computer code itself is a universal one in the sense that a user can give it any reaction descriptor, and it will interpret this descriptor as an operator which acts on any problem structure that he might wish to submit, and furthermore that it will accept any valence chemical boundary conditions that he might choose to impose. In short, chemical information is no more embedded in the code, but is only represented as pieces of data. The code only knows that the data have a chemical meaning, but it does not know the chemical information contained in the data.

1.3 Types of Problems

The DUGUNDJI-UGI-model (UII-792) (UII-793)(see section 2) is the basis for the deductive solution of problems in structural chemistry, but in its general, mathematically stated, form it is not directly usable as a tool for solving these problems in practice.

Two types of problems are of paramount importance in structural chemistry. The first of these is the application of a reaction to a given problem structure in all allowable ways: given a chemical structure and given a reaction, or a set of reactions, apply these reactions in all mathematically permitted and chemically meaningful ways to the problem structure. The result is a set of structures, which are the solution to this problem. This is the classical synthesis planning paradigm.

The second type of problem is documentation oriented: given a reaction equation, or a collection of reaction equations, together with more or less complete information about the problem structure, or set of problem structures, represent these reactions in such a way that it can be stored and retrieved with acceptable effort and that similar reactions are grouped together in the collection of data.

In practice, the second type of problem requires the definition of a canonical numbering and, possibly, a machine-readable representation of the reaction descriptor. The reaction descriptor in the DUGUNDJI-UGI-model would be an R-matrix together with associated parts of some BE-matrices. A machine-readable coding would take the form of strings with minimum redundancy.

2. THE DUGUNDJI-UGI MODEL

The DUGUNDJI-UGI-model was originally stated in mathematical terms (UII-731). The operators are represented by a matrix that is called the R-matrix. The operands are chemical structures which are coded in the form of matrices, called BE-matrices. The operation, mathematically, is a matrix addition. The master equation of the DUGUNDJI-UGI-model

$$^{1}\underline{B} + {}^{12}\underline{R} = {}^{2}\underline{B}$$

is a mathematically correct and chemically non-prejudiced way of representing a particular chemical reaction.

2.1 Bond-Electron Matrices

Chemical structures are represented by an atom vector, $\underline{a}$, which represents the atoms that are present in the structure, and the BE-matrix, $\underline{B}$, which represents the distribution of the valence electrons.

A chemical structure might be an individual molecule, as are the molecules 1, 2, 3 in the following three examples:

$\underline{1}$ = Carbon monoxide: $|\overset{2}{C} \equiv \overset{1}{O}|$

$$^{1}a = (O,C)$$

$$^{1}B = \begin{Bmatrix} 2 & 3 \\ 3 & 2 \end{Bmatrix}$$

$\underline{2}$ = Ammonia: $\overset{2}{H}-\overset{1}{\underline{N}}(-H^{3})-H^{4}$

$$^{2}a = (N, H, H, H)$$

$$^{2}B = \begin{Bmatrix} 2 & 1 & 1 & 1 \\ 1 & 0 & 0 & 0 \\ 1 & 0 & 0 & 0 \\ 1 & 0 & 0 & 0 \end{Bmatrix}$$

$\underline{3}$ = $\overset{4}{H}\ \overset{2}{C}\ \overset{1}{O}\ \overset{3}{N}\ \overset{5\ 6}{H_2}$

$$^{3}a = (O\ C\ N\ H\ H\ H)$$

$$^{3}B = \begin{Bmatrix} 4 & 2 & 0 & 0 & 0 & 0 \\ 2 & 0 & 1 & 1 & 0 & 0 \\ 0 & 1 & 2 & 0 & 1 & 1 \\ 0 & 1 & 0 & 0 & 0 & 0 \\ 0 & 0 & 1 & 0 & 0 & 0 \\ 0 & 0 & 1 & 0 & 0 & 0 \end{Bmatrix}$$

A chemical structure may also be an 'ensemble of molecules' (EM), as in the following example. Such an EM, typically, would be one side of a reaction equation. $\underline{4}$ is an EM, composed of the individual molecules $\underline{1}$ and $\underline{2}$:

$\underline{4}$ = $\overset{21}{CO}$ + $\overset{34,5,6}{NH_3}$ $(\underline{4} \equiv \underline{1} + \underline{2})$

$$^{4}a = (O\ C\ N\ H\ H\ H)$$

$$^{4}B = \begin{Bmatrix} 2 & 3 & 0 & 0 & 0 & 0 \\ 3 & 2 & 0 & 0 & 0 & 0 \\ 0 & 0 & 1 & 1 & 1 & 1 \\ 0 & 0 & 1 & 0 & 0 & 0 \\ 0 & 0 & 1 & 0 & 0 & 0 \\ 0 & 0 & 1 & 0 & 0 & 0 \end{Bmatrix}$$

We observe that, with a suitable numbering, the BE-matrix of an EM takes the form of a block matrix where each non-zero block is the BE-matrix of a molecule.

2.2 Reaction Matrices

While a BE-matrix represents the (static) distribution of the valence electrons within an EM, an R-matrix represents a redistribution of valence electrons that transforms one EM into another EM. R-matrices represent the dynamic aspect of structural chemistry.

The reaction $\underline{4} \longrightarrow \underline{3}$, i.e. $\underline{1} + \underline{2} \longrightarrow \underline{3}$, is represented by the equation:

$$CO + NH_3 \longrightarrow HCONH_2 \qquad (\underline{4} \longrightarrow \underline{3})$$

$$^{4}B + {}^{43}R = {}^{3}B$$

where:

$$^{43}R = \begin{array}{c} \begin{matrix} O & C & N & H & H & H \end{matrix} \\ \begin{Bmatrix} 2 & -1 & 0 & 0 & 0 & 0 \\ -1 & -2 & 1 & 1 & 0 & 0 \\ 0 & 1 & 0 & -1 & 0 & 0 \\ 0 & 1 & -1 & 0 & 0 & 0 \\ 0 & 0 & 0 & 0 & 0 & 0 \\ 0 & 0 & 0 & 0 & 0 & 0 \end{Bmatrix} \end{array}$$

The entries of an R-matrix have the following chemical meaning:

<u>negative</u> off-diagonal entries: bond breakages

<u>positive</u> off-diagonal entries: bond formations

<u>negative</u> diagonal entries: 'disappearing' free valence electrons

<u>positive</u> diagonal entries: 'new' free valence electrons

According to the law of conservation of matter, electrons cannot really appear or disappear; rather, their number remains constant, and thus the sum of the entries of an R-matrix must be zero.

Removal of all zero-rows and zero-columns from an R-matrix gives the 'irreducible R-matrix' which defines the 'reaction core' (BAJ-811). Reaction-invariant bonds within the reaction core are represented by the 'intact BE-matrix'. In the case of the reaction $\underline{4} \longrightarrow \underline{3}$, the intact BE-matrix would be:

$$^{43}{}_{a}I = (O \quad C \quad N \quad H)$$

$$^{43}{}_{B}I = \begin{Bmatrix} 2 & 2 & 0 & 0 \\ 2 & 0 & 0 & 0 \\ 0 & 0 & 2 & 0 \\ 0 & 0 & 0 & 0 \end{Bmatrix}$$

2.3 Reaction Generators

Although the R-matrices are operators, the master-equation of the DUGUNDJI-UGI-model (see section 2) is not directly implementable as a reaction generator, because the problem of a general reaction generator is to:

a) add a given R-matrix, $^{12}\underline{R}$, in all possible ways to the BE-matrix, $^{1}\underline{B}$, that represents the 'problem structure':

$$^{1}\underline{B} + {}^{k}\underline{P}\,{}^{12}\underline{R}\,{}^{k}\underline{P}^{-1} = {}^{k2}\underline{B}$$

where $^{k}\underline{P}$ is a member of the set of up to n! permutation matrices of rank n.

b) do this for <u>all</u> R-matrices that are of interest in your field of chemistry.

If we used the above matrix equation directly, the set of solutions, $^{k2}\underline{B}$, would contain an overwhelming number of chemically illegal solutions that would have to be sifted out.

In cases like this, there is a better approach than that of first evaluating all permutations and then sifting out the false drops later. In game theory, a comparable situation occurs: given a situation, evaluate all combinations of permitted moves. We know that every permutation of $^{12}\underline{R}$ is a legal R-matrix, and thus represents a 'sequence of permitted moves', mathematically speaking. But we also know that with the given problem structure, $^{1}\underline{B}$, many of these permutations would correspond to 'losing situations' of the chemical-structure game, i.e. they would lead to chemically illegal (or energetically unstable) structures. Such losing moves might be characterised by a violation of elementary constraints on BE-matrices; for example, elements of a BE-matrix would become negative ('negative' bond orders), i.e. we must not attempt to break a bond where there was no bond to begin with, or the application of a permuted R-matrix might be unacceptable because it would lead to a violation of the valence rules for a chemical element or for a substructure.

From game theory we learn that in such cases we should build up the permutations element by element. Every permutation appears as a node on a directed tree with a depth of n and an out-valence of n-d on each node at depth d. The strategy then consists in pruning all branches at each node that corresponds to an illegal move. Only those permutations are executed that successfully reach a terminal node, i.e. a leaf. This strategy was built into the code of the reaction generator of CICLOPS (UII-741), the first synthesis planning program on the basis of the DUGUNDJI-UGI-model, and of the successors of CICLOPS, e.g. EROS (GSJ-781;-791;-792). Since the strategy together with the reaction pattern was an integral part of the program code, a piece of program code had to be written for each individual R-matrix. This is clearly a rather cumbersome way of implementing a reaction generator. Nevertheless, CICLOPS and its offspring were relatively successful because the few reaction generators contained in them (GSJ-782) covered the reaction types that statistically are the most frequently occurring ones (BRJ-761) (BRJ-771) (BRJ-772) (BRJ-773), and because many more complex reaction patterns can be obtained by repeated execution of the simpler types.

However, the problem remains that these 'hard-wired' reaction generators do not give the user the freedom to really play the 'DUGUNDJI-UGI-game'. He cannot select a set of reaction patterns of his choice, and let the program evaluate these.

3. THE PATH-FINDING APPROACH

The solution to this problem was obtained through a deeper understanding of the formalism of R-matrices. At the beginning there was the problem of making the R-matrix a suitable vehicle for the description of reactions, so that it could be used in applications like the documentation of reactions.

3.1 R-Strings

If a matrix representation is to be useful as a descriptor and access key in conventional storage and retrieval systems, it must fulfill at least one condition: there has to be a canonical numbering. In practice, an additional condition has to be fulfilled: the matrix should be expanded into a string in such a way that non-relevant matrix elements are suppressed. We also wish that the string should have a semantic meaning, so that it represents the structure of the underlying problem.

We observe that most chemical reaction patterns contain paths of alternating bond formations and breakages. The commonly occurring metathesis reaction:

$$AB + CD \longrightarrow AD + BC$$

shows a four-membered circular alternating pattern:

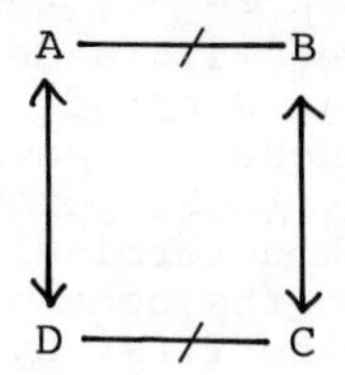

where the symbol ⟷ denotes a bond formation, and the symbol —/— a bond breakage (VEG-741). For simple cases like the one shown here, this observation was made earlier (AEJ-671) (AEJ-672) (AEJ-681) (AEJ-751) (AEJ-752) (AEJ-791) (AEJ-792) (AEJ-793). With a suitable numbering, the irreducible R-matrix that describes such a reaction pattern contains all the non-zero off-diagonal elements on the first side-diagonal (when the alternating path is open), and one additional non-zero element in the upper right (and lower left) corner (when the path is circular). Reading the side-diagonal- and corner-elements in sequence yields a convenient string notation ('R-string') for such simple R-matrices. For brevity, we make the alternating sign implicit. The string is enclosed in parentheses. The R-string for this case, then, is: (1111).

More complicated reaction patterns also contain an alternating path, as with the reaction $\underline{4} \longrightarrow \underline{3}$:

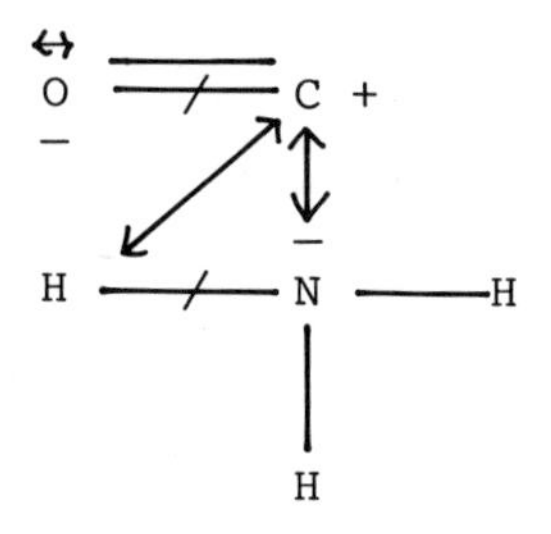

Here, the higher side-diagonals also contain non-zero elements, and we append these, separated by a semicolon, to the first side-diagonal. Trailing zeros are omitted while extraordinary reversals of sign are indicated by a full stop preceding the number. When the main diagonal contains non-zero elements, it is appended to the string of the side diagonals, separated by a slash. The sign is not implicit here, being always given explicitly by a full stop. Thus the reaction $\underline{4} \longrightarrow \underline{3}$ yields the string: (1110;01/2.2).

A canonicalisation rule can now be established by postulating the lexicographically greatest R-string (BAJ-811) (BAJ-812).

The intact BE-matrix, then, is expanded in the same manner as the R-matrix along the side-diagonals, but, of course, without implicit signs which would be meaningless here, because a BE-matrix can only contain positive elements.

Finally the atom vector of the reaction core is added, separated with a colon, to give a full string representation of the reaction. The concept of an alternating path proved to be crucial in more than one respect since it allowed the implementation of a very fast and efficient canonicalisation algorithm (BAJ-831), which was developed by Annette von Scholley (SHA-811) at T.U. München before she joined the group of Professor M. Lynch at the University of Sheffield.

3.2 Valence Schemes

Valence schemes are employed to implement a further aspect of structural chemistry, namely the chemically permitted distribution of the valence electrons around an atom. A valence scheme (BAJ-781) (BAJ-801) BAJ-811) is a vector, $\underline{v}$. The i-th element of $\underline{v}$ denotes the number of bonds with a bond order of i, and the 0-th element denotes the number of free electrons around a given atom. This is illustrated by the following examples:

$\underline{v}$ = (0,4,0,0,...) four single bonds, as with aliphatic carbon, silicon, etc.

$\underline{v}$ = (0,2,1,0,...) two single bonds and one double bond, as with olefinic carbon

$\underline{v}$ = (4,2,0,0,...) four free electrons and two single bonds, as with oxygen in alcohols, ethers

$\underline{v}$ = (4,0,1,0,...) four free electrons and two single bonds, as with oxygen in ketones, aldehydes

$\underline{v}$ = (0,1,0,1,...) one single bond and one triple bond, as with acetylenic carbon

$\underline{v}$ = (1,3,0,0,...) one free electron and three single bonds, as with aliphatic carbon free radicals

We note that, in general, each chemical element will command more than one valence scheme and that one valence scheme can apply to various chemical elements. Tables of valence schemes and corresponding chemical elements are supplied to the reaction generator as data files. By selecting (or manipulating) these data files the user can select those areas of chemistry that he wants to explore.

3.3 Implementation of the Reaction Generator

The concept of the alternating path also led to a novel approach for writing a universal reaction generator (SAK-841).

The notion of a path and the task of finding all allowed mappings of a given path onto a given molecular graph suggests the use of a recursive approach. This type of problem is often found in graph theory, and we note that the transformation of the problem from its matrix formulation into a suitable string formulation is the clue to the solution, because the string represents a path of alternate bond breakages and formations in most cases.

The R-string is evaluated by a recursive PASCAL program. Depending on the value of the reaction string, the program scans those atoms of the problem structure that are as yet unused, for possible bond formations, or the neighbours of the atom in use for possible bond breakages. In effect, it constructs all possible paths of bond breakages and formations across the graph that represents the problem structure which are in accordance with the R-string.

The algorithm may be sketched as follows:

```
Initialise:
<atom_in_use>   :=  0
<set_of_used>   :=  empty

Recursion:
<atom_in_use>   := next form of atoms not in <set_of_used>
case <R-string> of
<bond formation>: tentatively make ...  } ... a bond to all
                                        } other atoms
<bond breakages>: tentatively break ... } not in <set_of_used>

<no action>:      do nothing;
esac;

if  <check_of_valence_scheme>
    then  <set_of_used> :=  <atom_in_use>  + <set_of_used>
        Recursion;

    neht
    else reset tentative bond breakage/formation
fi

Termination:
if R-string exhausted
    then Output_Reaction_Product_Structure;
    return;
    neht
```

What this strategy has in common with the approach mentioned in section 2.3 is that the solution is built up in steps and that all unsuccessful branches are pruned immediately. But now the reaction operator itself is not part of the code: only the general strategy of executing a step and checking its validity in a given context before executing the next step or marking a successful solution is part of the code. The reaction operator itself is a piece of data.

These data may either be provided as a file of common or problem-oriented special reaction patterns, or they may be entered manually by an inquisitive user who looks for the unusual.

5. SOME ILLUSTRATIVE RESULTS

A sample run may illustrate the scope of results that can be obtained, ranging from 'textbook chemistry' to highly speculative results. The simple reaction pattern AB + CD ⟶ AD + BC, which statistically accounts for more than two thirds of all known reaction patterns (BRJ-761) (BRJ-771) (BRJ-773), was run against the EM consisting of acetone and hydrogen cyanide.

It should be noted that all reactions can be read in the reverse sense.

5.1 Intramolecular Arrangements and Fissions of One Reactant

The restriction to non-intramolecular reactions (which is a simple set-operation in PASCAL) was relaxed for this run, so that reactions without participation of the other reactant are also produced. They range from a null-reaction (Product 67) to interconversion between acetone and its enol (Product 1), oxirane (Product 7) and aldehyde (Product 33); in addition, structures of high-energy products like cyclopropanone (Product 70) or ketene (Product 36) are generated. For these, the reverse reaction would, of course, be more likely under normal conditions.

Product 67

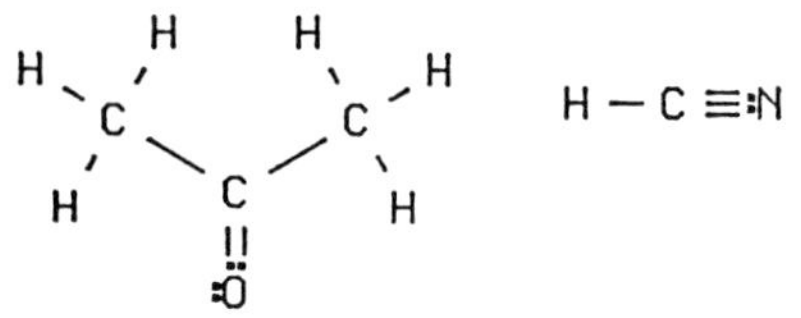

Product 1

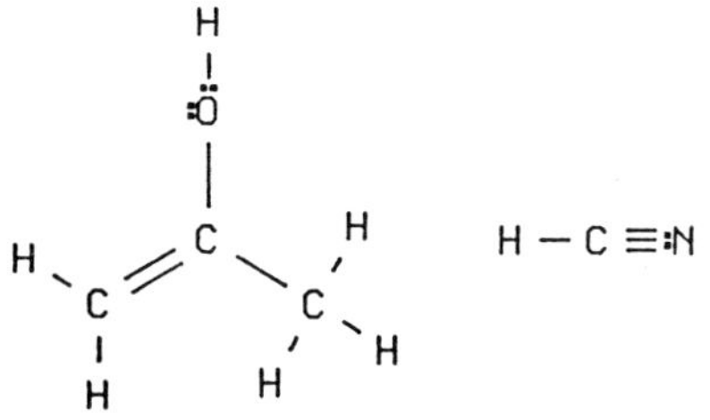

Product 7

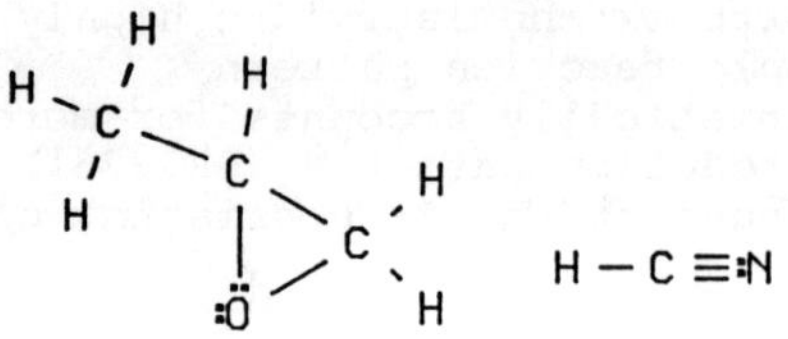

H — C ≡ N

Product 33

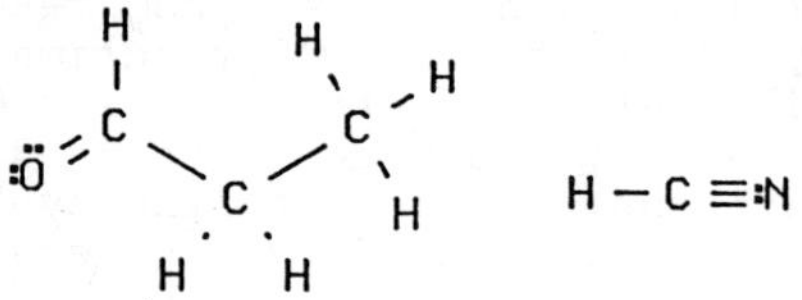

H — C ≡ N

Product 70

C = O H — C ≡ N H — H

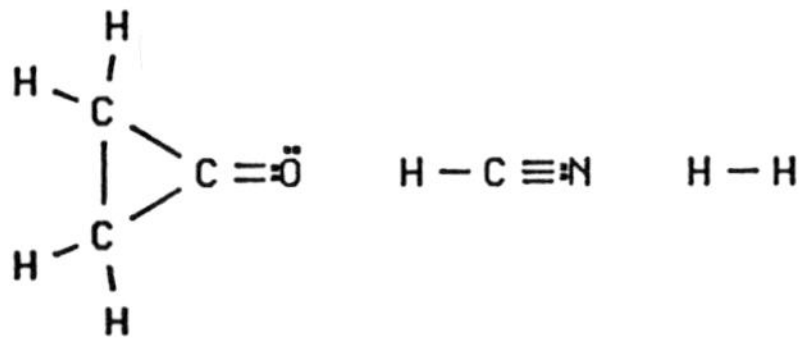

Product 36

O = C = C H — C ≡ N

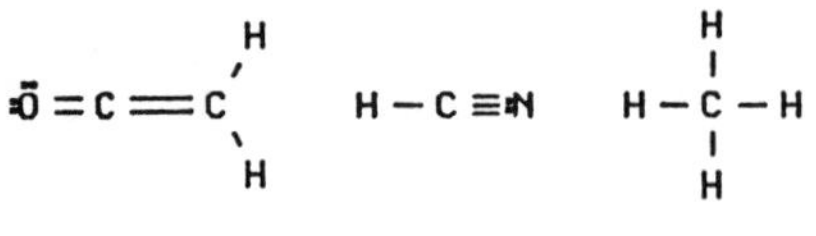

5.2 Condensations and Intermolecular Rearrangements

The structure of the 'textbook-product' cyanhydrin (Product 15) is produced, as well as more speculative condensation products (Product 16, Product 39, Product 40, Product 73, Product 74). Among these, some four-membered structures (Product 13, Product 14) look quite interesting. Some of the reactions involve breaking and forming of C-C and C-H bonds only (Product 41, Product 42, Product 76).

Product 15

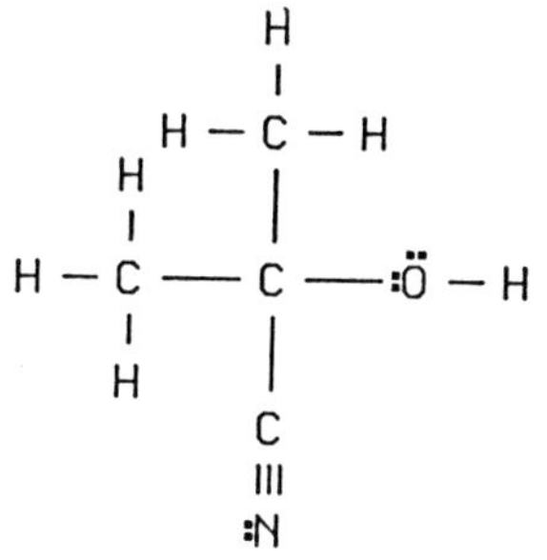

```
            H
            |
        H - C - H
     H      |
     |      |
 H - C ---- C ---- :Ö - H
     |      |
     H      |
            C
            |||
           :N
```

Product 16

```
                   H
                   /
           H  -  C  -  H
                 /
                /    H
               C  -
     :Ö  ---         H
    /          \   .
:N ≡ C           C
               .   \
             H       H
```

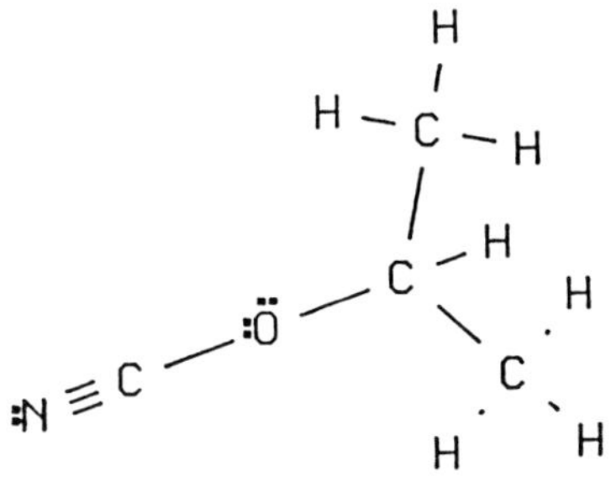

Product 39

```
                    :Ö
      H             \\          H
      /              C  -----   /
  H - C ---- :N ==  /           C - H
      /            C            /
      H             \           H
                     H
```

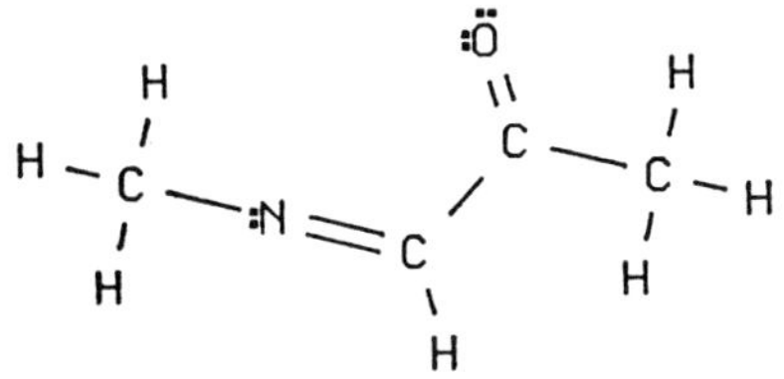

Product 42

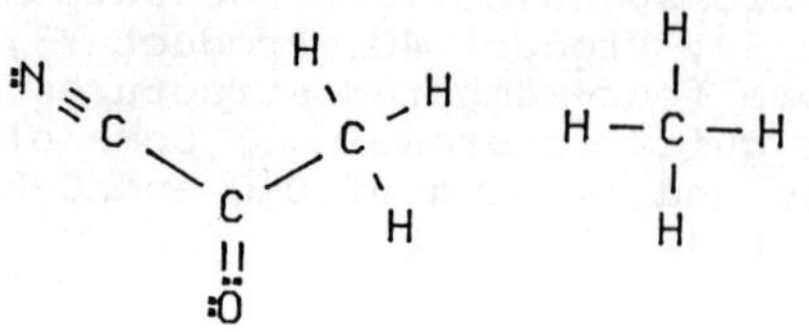

Product 76

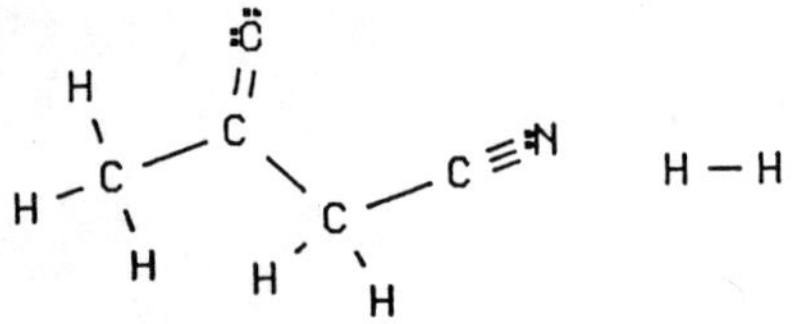

Product 41

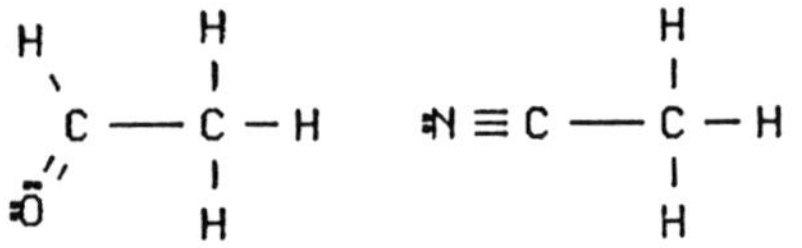

Product 40

Product 73

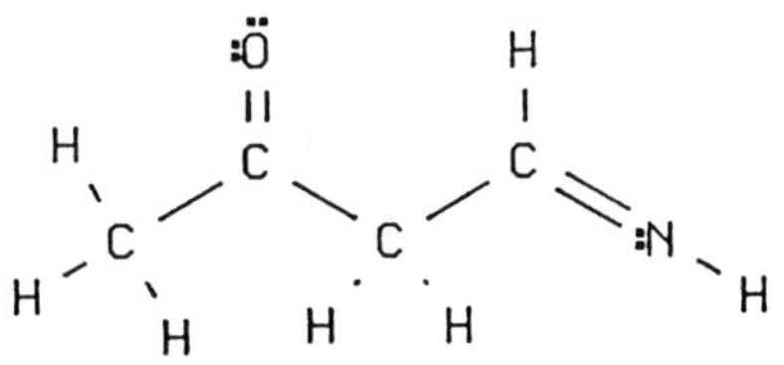

Product 74

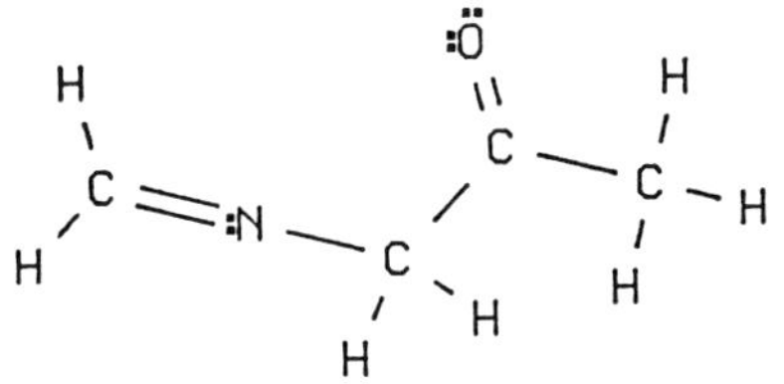

Product 13

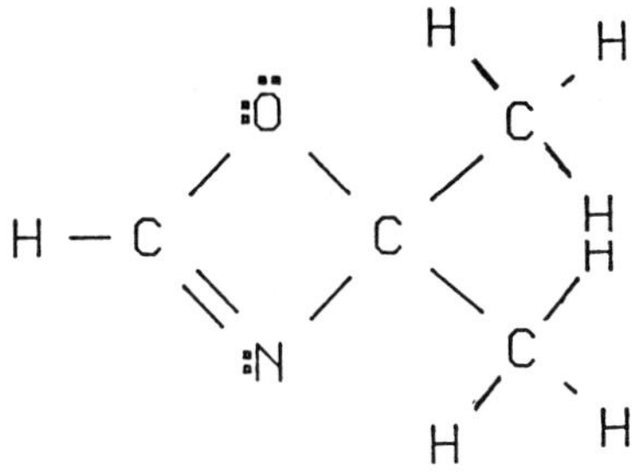

Product 14

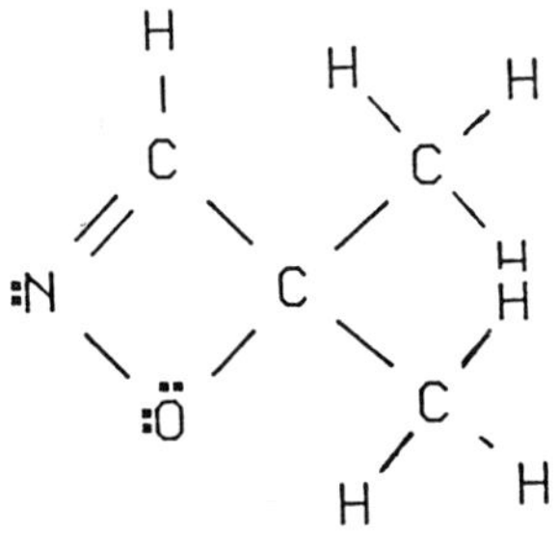

6. CONCLUSION

The merit of a reaction generator in a reaction documentation system is illustrated by a run of the aforementioned reaction string (1110;01/2.2) against the EM 3. When the hydride pattern is included among the valence schemes for hydrogen, then not only is the formamide generated, but also a cyclic structure: Product 7.

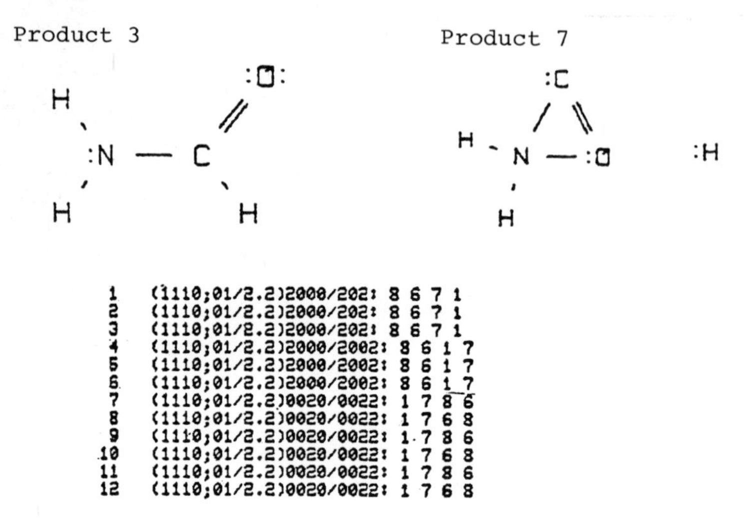


```
1   (1110;01/2.2)2000/202: 8 6 7 1
2   (1110;01/2.2)2000/202: 8 6 7 1
3   (1110;01/2.2)2000/202: 8 6 7 1
4   (1110;01/2.2)2000/2002: 8 6 1 7
5   (1110;01/2.2)2000/2002: 8 6 1 7
6   (1110;01/2.2)2000/2002: 8 6 1 7
7   (1110;01/2.2)0020/0022: 1 7 8 6
8   (1110;01/2.2)0020/0022: 1 7 6 8
9   (1110;01/2.2)0020/0022: 1 7 8 6
10  (1110;01/2.2)0020/0022: 1 7 6 8
11  (1110;01/2.2)0020/0022: 1 7 8 6
12  (1110;01/2.2)0020/0022: 1 7 6 8
```

Along with the product structure, the program outputs a complete description of the reaction in string notation, giving the R-string, the string notation for the intact BE-matrix and the atom vector of the reaction core. These may be used directly as a query to the reaction documentation system (BAJ-791) (BAJ-801) (BAJ-841) that was described at the CSA meeting in Exeter in 1982 (ASJ-851). The duplicates (1 ... 6, 7 ... 12) are generated in a non-canonical numbering: when used as a query to the reaction documentation system, they would be recognised and eliminated because the reaction documentation system canonicalises every query before processing it.

The work has shown that the inclusion of a reaction generator in the front end of a reaction documentation system would provide a significant enhancement to the power of such a system and give it some of the capabilities of an expert system.

Acknowledgement

This work was supported in part by the Commission of the European Community under Contract no. ENV-677-D (B).

REFERENCES

AEJ-671: J. F. Arens, Chem. Weekbl., 63, 513, 1967.

AEJ-672: J. F. Arens, Chem. Weekbl., 63, 529, 1967.

AEJ-681: J. F. Arens, 'Tele-reactions', Bull. Soc . Chim. Fr., 1968, 3037-3044, 1968.

AEJ-751: J. F. Arens, 'Eenheid in verscheidenheid' (Voordracht 25. Oct. 1975) Versl. der gew. vergaderingen der afd. Natuurkunde, Kon. Ned. Akad. van Wetanschappen, 84, 155-166, 1975.

AEJ-752: J. F. Arens, 'Interrelations of unsaturated aliphatic compounds. I. General considerations', Recl. Trav. Chim. Pays-Bas, 94, 3-8, 1975.

AEJ-791: J. F. Arens, 'A formalism for the classification and design of organic reactions. I. The class of $(-+)_n$ reactions', Recl. Trav. Chim. Pays-Bas, 98, 155-161, 1979.

AEJ-792: J. F. Arens, 'A formalism for the classification and design of organic reactions. II. The classes of $(+-)_n+$ and $(-+)_n-$ reactions', Recl. Trav. Chim. Pays-Bas, 98, 395-399, 1979.

AEJ-793: J. F. Arens, 'A formalism for the classification and design of organic reactions. III. The class of $(+-)_nC$ reactions', Recl. Trav. Chim. Pays-Bas, 98, 471-500, 1979.

ASJ-851: J. Ash, P. Chubb, S. Ward, S. Welford, P. Willett, 'Communication, storage and retrieval of chemical information', Chichester, Ellis Horwood, 1985.

BAJ-791: J. Brandt, 'Hierarchische Klassifizierung der Umwandlungen von Systemen binärer Relationen und deren Anwendung aug chemische Reaktionen', Bund. Min. Forschung und Technologie (Bonn) und Gesellschaft für Information und Dokumentation (Frankfurt). Projekt PT 63.301. Berichte (1978 ... 1980).

BAJ-801: J. Brandt, 'Hierarchisch strukturierte Speicherung und Ermittlung von chemischen Reaktionen', Bund. Min. Forschung und Technologied (Bonn) und Gesellschaft für Information und Dokumentation (Frankfurt). Projekt PT 63.302. Berichte (1980 ff.)

BAJ-781: J. Brandt, J. Friedrich, J. Gasteiger, C. Jochum, W. Schubert, P. Lemmen, I. Ugi, 'The deductive solution of chemical problems by computer programs on the basis of a mathematical model of chemistry', Pure Appl. Chem., 50, 1303-1318, 1978.

BAJ-811: J. Brandt, 'Ein mathematisch begründetes hierarchisches Ordnungssystem chemischer Reaktionen und dessen theoretische und praktische Anwendungen', Habilitationsschrift, TU-München, 1981.

BAJ-812: J. Brandt, J. Bauer, R. M. Frank, A. v. Scholley, 'Classification of reactions by electron shift patterns', Chemica Scripta, 18, 53-60, 1981.

BAJ-831: J. Brandt, A. v. Scholley, 'An efficient algorithm for the computation of the canonical numbering of reaction matrices', Computers & Chemistry, 7, 51-59, 1983.

BAJ-841: J. Brandt, A. v. Scholley, M. Wochner, K. Stadler, 'A documentation system for chemical reactions' 188th ACS National Meeting, Division of Chemical Information, Symposium Chemical Reaction Data Bases, 28-31.8.1984.

BRJ-761: J. C. J. Bart, E. Garagnani, 'Organic reaction schemes and general reaction-matrix types, I. Rearrangement reactions', Z. Naturforsch., B31, 1646-1653, 1976.

BRJ-771: J. C. J. Bart, E. Garagnani, 'Organic reaction schemes and general reaction-matrix types, II. Basic types of synthetic transformations', Z. Naturforsch., B32, 455-464, 1977.

BRJ-772: E. Garagnani, J. C. J. Bart, 'Organic reaction schemes and general reaction-matrix types, III. A quantitative analysis', Z. Naturforsch., B32, 465-468, 1977.

BRJ-773: J. C. J. Bart, E. Garagnani, "Organic reaction schemes and general reaction-matrix types, IV. Organic name reactions', Z. Naturforsch., B32, 678-683, 1977.

GSJ-781: J. Gasteiger, C. Jochum, 'EROS, a computer program for generating sequences of reactions', Top. Curr. Chem., 74, 93-126, 1978.

GSJ-782: J. Gasteiger, 'Programmsystem EROS, Benutzer-Manual', Bericht, 07.01.1978.

GSJ-791: J. Gasteiger, C. Jochum, M. Marsili, J. Thoma, 'Das Syntheseplanungsprogramm EROS', Informal Commun. Math. Chem., 6, 177-199, 1979.

GSJ-792: J. Gasteiger, C. Jochum, M. Marsili, J. Thoma, 'The synthetic design program EROS' in: (KPV-791) pp. 492-457.

HDE-742: W. T. Wipke, S. R. Heller, R. J. Feldmann, E. Hyde, 'Computer representation and manipulation of chemical information', Wiley, New York, 1974, pp.105ff. ISBN 0-471-95595-7

KPV-791: V. A. Koptyug (ed.) 'Doklady predstavlennye na IV. mezhdunarodnuyu konferentsiyu po primeneniyu EVM v khimii (II) 19-25 iyunya 1978', Novosibirskii Inst. Organ. Khimii SO AN SSSR Sib. Otdel. Akad. Nauk SSSR, Novosibirsk, 1979.

SAK-841: K. Stadler, 'Ein parametrisierter Reaktionsgenerator auf der Basis des DUGUNDJI-UGI-Modells', Diplomarbeit, T.U. München, 1984.

SHA-811: A. von Scholley, 'Hierarchische Klassifizierung und Dokumentation von chemischen Reaktionen', Dissertation, T.U. München, 1981.

UII-731: J. Dugundji, I. Ugi, 'An algebraic model of constitutional chemistry as a basis for chemical computer programs', Topics Curr. Chem., 39, 19-64, 1973.

UII-741: J. Blair, J. Gasteiger, C. Gillespie, P. D. Gillespie, I. Ugi, 'CICLOPS - a computer program for the design of syntheses on the basis of a mathematical model', in: (HDE-742) pp.129-146.

UII-792: I. Ugi, J. Bauer, J. Brandt, J. Friedrich, J. Gasteiger, C. Jochum, W. Schubert, 'Neue Anwendungsgebiete für Computer in der Chemie', Angew. Chemie, 91, 99-111, 1979.

UII-793: I. Ugi, J. Bauer, J. Brandt, J. Friedrich, J. Gasteiger, C. Jochum, W. Schubert, 'New applications for computers in chemistry', Angew. Chemie Int. Ed., 18, 111-123, 1979.

VEG-741: G. E. Vleduts, E. A. Geivandov, 'Abtomatizirovannye Informatsionnye Sistemy dlya Khimii', Nauka, Moskva, 1974. (C.A. 81, 12584d)

17 Compound - oriented and reaction - oriented structural languages for reaction data base management

J.-E. Dubois, G. Sicouri and R. Picchiottino
Institute de Topologie et de Dynamique des Systèmes, Université Paris VII

ABSTRACT

In the DARC (Description, Acquisition, Retrieval, Computer-aided design) System context, two reaction management languages are compared in a simplified pedagogical form. The compound-oriented language handles pairs of initial-final structures, while the reaction-oriented language handles transformations of an initial structure into a final structure. A transformation is defined as localisation on an invariant of ablation and adjunction variations (IGLOO [Invariant Graph and Localised Ordered Operators] method). For a single reaction both models have totally equivalent informational content, and transcoding is feasible. The main advantage of the compound-oriented model is the direct reflection of available experimental observation whenever a reaction is carried out. The reaction-oriented model is motivated by conciseness and significant grouping. Conciseness is necessary when a whole data base of reactions has to be stored. Grouping is needed for substructural searching, and for a generic description of structurally related reactions. These discrepancies in efficiency stem from the differences in nature and the differences in scope of associated reaction-substructure relationships. Transformation-substructures are more precise than pairs of usual compound-substructures. Moreover they can be specifically parametrised according to transformational constraints, i.e. related to localisation, variations or invariant sites. Use of transformation-substructures to reduce noise resulting from a classical query via a pair of substructures is illustrated on the KETO-REACT reaction data base. Examples of access to classes of ketone preparation methods are given.

At the confluence of structural and mechanistic chemistry on one hand, computerised treatment of graphs and their transformations on the other hand (Figure 1), reaction management is a very exciting challenge.

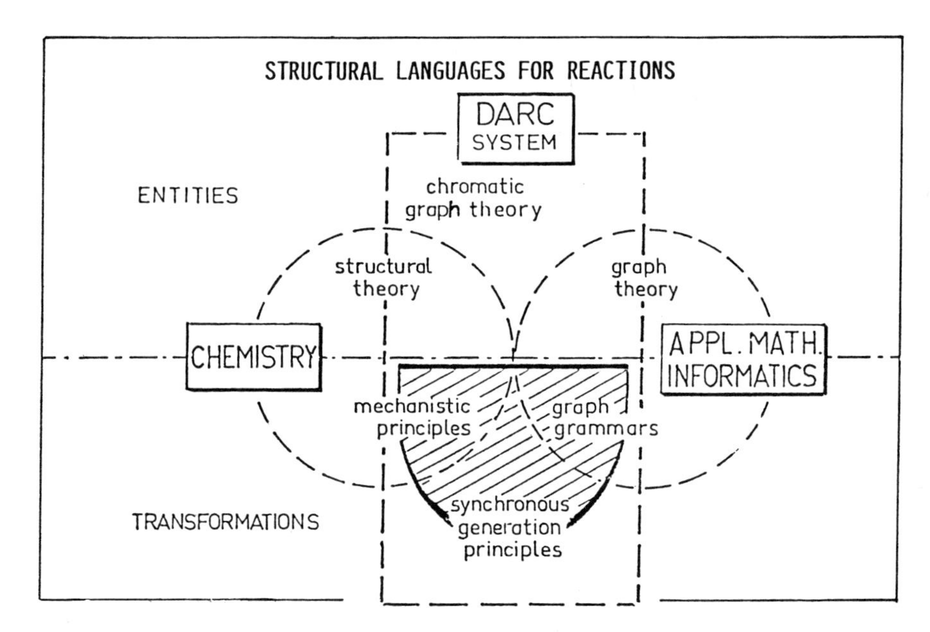

Figure 1.

In a recent review article (1), G. Gello observed that, compared to the (relatively) comfortable situation for compound handling, we are still in a 'naive' period for automatic treatment of chemical reactions. The reasons given for the scarcity and relative lack of success of Reaction Management Systems are: a) 'the lack of a real and efficient methodology for analysing the theoretical aspects of reactions' and b) 'the necessity of translating them from usual languages into codes'. The first criticism is directly linked to the chemist's still poor understanding of intimate reaction mechanisms: actually only very few mechanisms are reliably established and several hypothetical mechanisms are often proposed for a single reaction. The second criticism concerning adequate description language of reaction is closely related to the first one: although improved computer techniques now exist for the translation of chemical formulas and associated text into internal machine representation, obviously this cannot compensate for the lack of a precise general structural modelling of reactions.

In existing reaction Data Base Management Systems (e.g. KETO-REACT (2)) logical schemes essentially rely on a Compound-Oriented (C-O) model: basically a reaction is considered as a set of initial compound structures, plus a set of final compound structures. In a Reaction-Oriented (R-O) model dynamic transformational aspects of a reaction are taken into account. R-O models have mainly been used for generic retro-reactions in synthesis planning (3), and in systematic reaction type classification works (4), but not for specific reaction data banks.

The aim of this paper is to examine which modelling option is more efficient faced with the methodological and practical criticisms above. C-O and R-O languages are thus compared - at structural and substructural level - in particular in view of their use in reaction Data Base substructural search and retrieval.

For the sake of clarity, we restrict our presentation to some simple classes of compounds (no charges, no abnormal valencies, no stereochemistry), of compound substructures, (no free sites (5), no loose chromaticities (6), no pending attachments (7), no generic constraints) and of reactions (15) (no stoichiometric coefficients, no parasite reactions). Indeed a mathematical formalism is necessary to highlight some conceptual differences. This theoretical effort is justified by important consequences in practical reaction indexing and retrieval.

1. TWO STRUCTURAL MODELLING OPTIONS FOR REACTIONS

Fundamental differences existing in two modes for representing structural information of a reaction are illustrated hereafter.

1.1 Pair of structures

The first option for structural modelling of a chemical reaction relies on an external conception of the reaction, considered as a black box: it only takes into account the structure of initial compounds, and the structure of final ones. To define the model we need to recall the chromatic graph concept (8):

DEFINITION - A chromatic graph is a quadruplet $G = (X, U, \chi_{NA}, \chi_{LI})$ where: X is the set of nodes; U is the set of edges, a subset of $A(X) = \{ \{x,y\} \mid x,y \in X, x \neq y \}$; χ_{NA} is the chromatic node function, an application of X in K_{NA} table of chemical elements; χ_{LI} is the chromatic edge function, an application of U in K_{LI} table of chemical bond types.

A chemical structure is derived from such a chromatic graph by taking into account valency constraint (Table 1a), and by using an isomorphism relationship:

DEFINITION - Two chromatic graphs G and G' are isomorphic if there is a bijective mapping of X onto X', that preserves the edges, and the chromaticities of nodes as well as of edges.

Since isomorphism is an equivalence relationship, chromatic graphs can be grouped into equivalence classes:

DEFINITION - A (chemical) structure S is a class of iso-

(a) for all $x \in X$,

$$\Delta V(x) = -V(\chi_{NA}(x)) + \sum_{\{x,y\} \in U} M(\chi_{LI}(\{x,y\})) \leqslant 0$$

where :

$V(\chi_{NA}(x))$ denotes the valency of the chemical element $\chi_{NA}(x)$ associated with node x ;

$M(\chi_{LI}(\{x,y\})$ denotes the formal multiplicity of the bond type $\chi_{LI}(\{x,y\})$ associated with edge $\{x,y\}$.

(b) 1) for $c = i, e_1, e_2$ and for all $x \in X_c$,

$$\Delta V_c(x) = -V(\chi_{NA}(x)) + \sum_{\{x,y\} \in U_e} M(\chi_{LI}(\{x,y\})) \leqslant 0$$

2) for all $x \in X_\ell$

$$\Delta V_\ell(x) = -V(\chi_{NA}(x)) + \max_{k \in \{1,2\}} \Big(\sum_{\{x,y\} \in U_\ell} M(\chi^s_{\Gamma LIk}(\{x,y\}) + \sum_{\{x,y\} \in U_{ek}} M(\chi_{LI}(\{x,y\}))\Big) + \sum_{\{x,y\} \in U_i} M(\chi_{LI}(\{x,y\})) \leqslant 0$$

where :

for $k \in \{1,2\}$, if $\chi_{\Gamma LIk}(u) = (k_\ell, k_s)$, $\chi^s_{\Gamma LIk}(u) = k_s$

Table 1: Valency constraints on a chromatic graph (a) and on a chromatic transfograph (b).

morphic chromatic graphs respecting the valency constraint.

A reaction R whose initial structure is S_1 and final structure S_2 is then simply modelled as a pair of structures $P = (S_1, S_2)$, represented by a pair of chromatic graphs (G_1, G_2).

1.2 Transfostructure

The second option for structural modelling of a chemical reaction involves access to the inner dynamics of the reaction, by enhancing common and distinguishing parts of the initial structure and the final structure. We define the model beginning with the chromatic transfograph concept (9) (Figure 2).

DEFINITION: A chromatic transfograph is a quadruplet $F = (X, U, \chi_N, \chi_L)$ where:

- $X = (X_i, X_\ell, (X_{e1}, X_{e2}))$ is composed of four disjoint sets of nodes called, internal, loose, and external (1 for ablation, and 2 for adjunction);

- $U = (U_i, U_\ell, (U_{e1}, U_{e2}))$ is composed of four sets of edges called internal, loose, and external (1 for ablation and 2 for adjunction); U_i is a subset of $A(X_\ell \cup X_i)$, U_ℓ a subset of $A(X_\ell)$, U_{e1} a subset of $A(X_\ell \cup X_{e1})$, and U_{e2} a subset of $A(X_\ell \cup X_{e2})$; every loose node belongs to an ablation or an adjunction edge;

- χ_N is composed of the sole function χ_{NA}; it is an application of $X_i \cup X_\ell \cup X_{e1} \cup X_{e2}$ in K_{NA};

- $\chi_L = (\chi_{LI}, \chi_{LI\ell}, (\chi_{\Gamma LI1}, \chi_{\Gamma LI2}))$ is composed of χ_{LI} the bond type function, $\chi_{LI\ell}$ the bond type looseness function, $\chi_{\Gamma LI1}$ and $\chi_{\Gamma LI2}$ the ablation and adjunction bond type specification functions; χ_{LI} is an application of $U_i \cup U_{e1} \cup U_{e2}$ in table K_{LI}, $\chi_{LI\ell}$ of U_ℓ in $K_{LI\ell}$ looseness table, $\chi_{\Gamma LI1}$ and $\chi_{\Gamma LI2}$ of U_ℓ in $K_{\Gamma LI}$ specification table; $K_{LI\ell} = \{\{k_1, k_2\} \mid k_1, k_2 \in K_{LI}, k_1 \neq k_2\}$ and $K_{\Gamma LI} = \{\{k_\ell, k_s\} \mid k_\ell \in K_{LI\ell}, k_s \in k_\ell\}$.

A transformation structure, or 'transfostructure' is derived from such a chromatic transfograph by taking into account valency constraints (Table 1b), and by using an isomorphism relationship (Figure 3):

DEFINITION - Two chromatic transfographs F and F' are isomorphic if there is a bijective mapping of X onto X', that preserves the node character (i, ℓ, e1, e2) and chromaticity, the edges, their character and their chromaticity.

Since isomorphism is an equivalence relationship, chromatic transfographs can be grouped into equivalence classes:

DEFINITION - A transfostructure T is a class of isomorphic chromatic transfographs respecting the valency constraint (Table 1b).

Such a transfostructure T displays the elements marking the transformation from an initial into a final structure: the invariant regroups loose and internal items, the ablation variation is composed of external-1 items, and the adjunction variation of external-2 items; loose items indicate the localisation of variations on the invariant (Figure 4).

A chemical reaction can then be modelled by a whole population of transfostructures. Each T corresponds to a specific 'rough' mechanism depicted by broken and newly formed bonds, with their localisation on variation sites. Any transfograph from T can be displayed using an original graphic notation system based on arrows pointing to or away from the bond (14) (Figure 4).

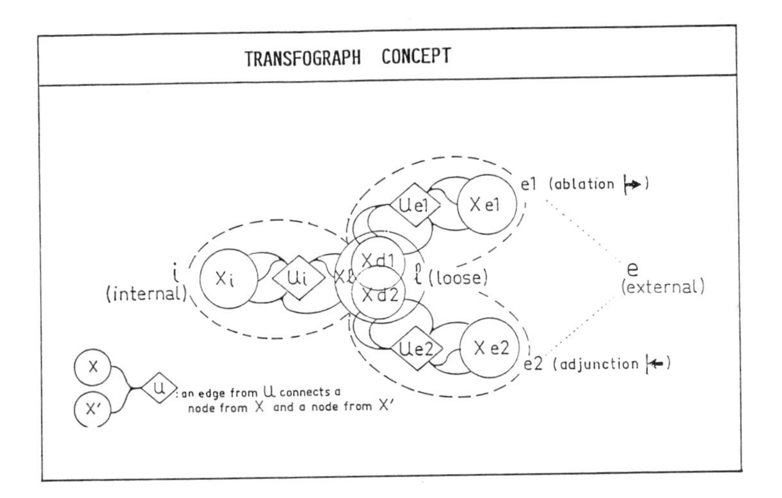

Figure 2.

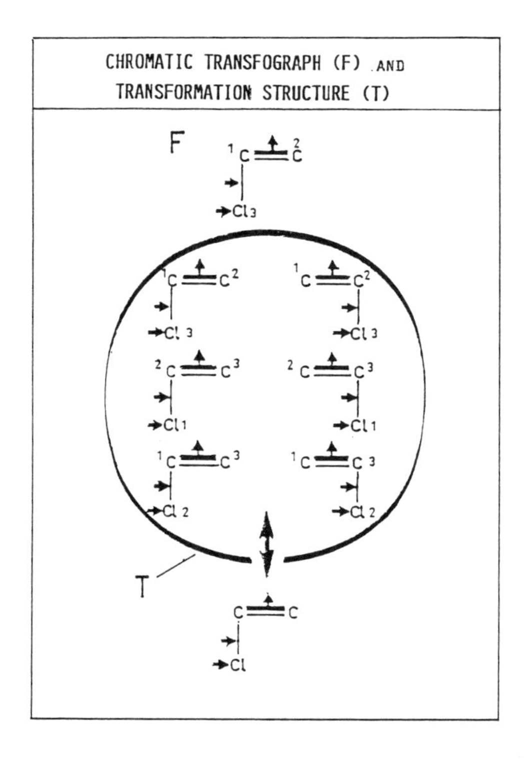

Figure 3.

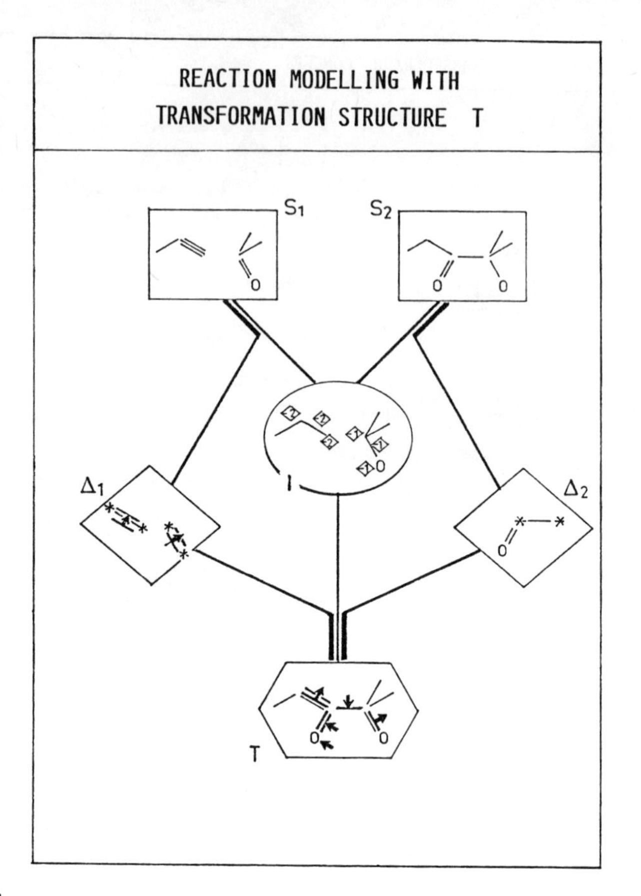

Figure 4.

1.3 Bridging the two models

We examine successively the possibility of deriving pair P from transfostructure T, and the reciprocal building of T from P.

The pair (G_1,G_2) is easily derived from transfograph F: G_1 is constructed from loose, internal and ablation items, and G_2 from loose, internal and adjunction items. This property is maintained by isomorphism: let (G_1,G_2) and (G_1',G_2') be two pairs of graphs derived from two transfographs F and F' both belonging to transfostructure T; G_1 and G_1' are isomorphic and belong to the same structure S_1, and in the same way G_2 and G_2' belong to S_2. Thus $P = (S_1,S_2)$ is univocally derived from T (10) (Figure 5).

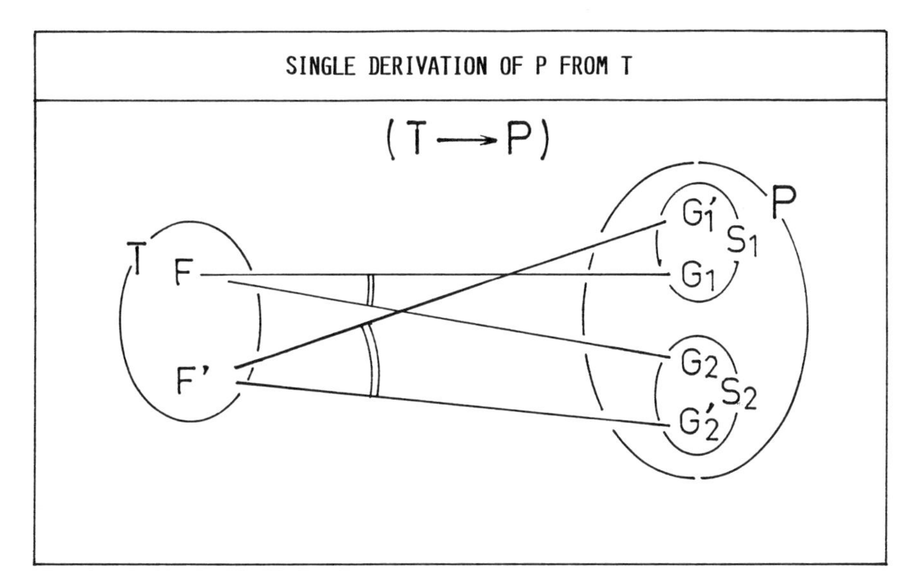

Figure 5.

The reciprocal building of T from P is less straightforward. Internal and loose items of transfograph F are constructed from a partial chromatic subgraph common to G_1 and G_2; the set of external-1 items is specific to G_1 and connected to loose items, and in the same way the set of external-2 items is specific to G_2 and connected to loose items. Two pairs (G_1,G_2) and (G_1',G_2'), with G_1,G_1' belonging to S_1 and G_2, G_2' belonging to S_2, generally result in two non isomorphic transfographs F and F'. This ambiguity is due to quantitative and qualitative recovering problems: size of common subgraphs can be variable, and even with equal size, their nature and embedding can differ. Thus from a given $P = (S_1,S_2)$ several T transfostructures can be constructed (Figure 6).

1.4 Unique modelling problems

Whereas the $P = (S_1,S_2)$ model for reaction $(S_1 \longrightarrow S_2)$ is naturally unique, there are different ways to choose a unique T model: 1) arbitrary optimisation criteria (purely formal approach), 2) evidenced mechanistic information (chemical approach), and 3) formal expression of general mechanistic or classification rules. The DARC-IGLOO method (11) makes successive choices relying on the latter type of formal expressions:

1) determination of the invariant (6) I as a population of Greatest Common SubStructures (GCSS) between S_1 and S_2 (Figure 7);

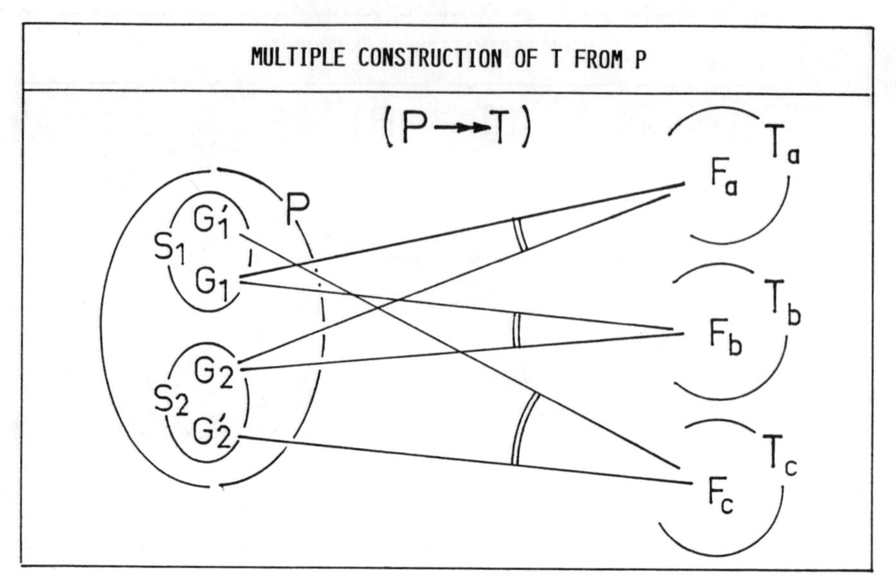

Figure 6.

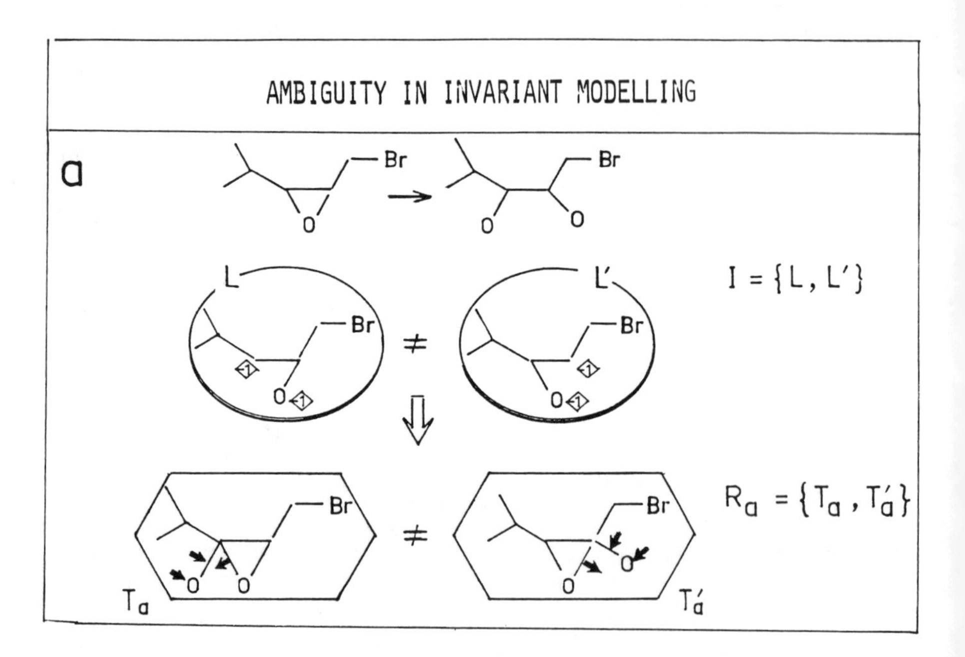

Figure 7.

2) determination of a population Δ of possible pairs of ablation and adjunction variations (10) as minimal attachment co-structures of GCSS in S_1 and S_2 (Figure 8);

3) building of T transfostructures by connection (9) between GCSS from I and corresponding co-structures in Δ (Figure 9);

4) total ordering of all candidate transfostructures according to transfograph generation elementary adjunction rules (10).

At each of the first three steps, ambiguity can theoretically arise, but actually happens very seldom (less than 5% on a 3000 reaction sample analysed in ketone protection Reaction Data Base PRO-REACT (12)). Thus we rarely have recourse to the purely formal fourth step.

At this stage, we note that C-O model (characterised by pair P) and R-O model (characterised by transfostructures T) have, in the absolute, totally equivalent information contents, while being presented in different forms. According to the goal, the passage from one model to the other will induce an added value. The main advantage of the C-O model is the direct reflection of information always available when a reaction is carried out experimentally. The R-O model is closer to mechanistic reality, and leads to a more compact representation of the reaction because it reduces structural information redundancy between initial and final structures. Real motivation of the R-O model will arise when significant grouping is sought (see next section).

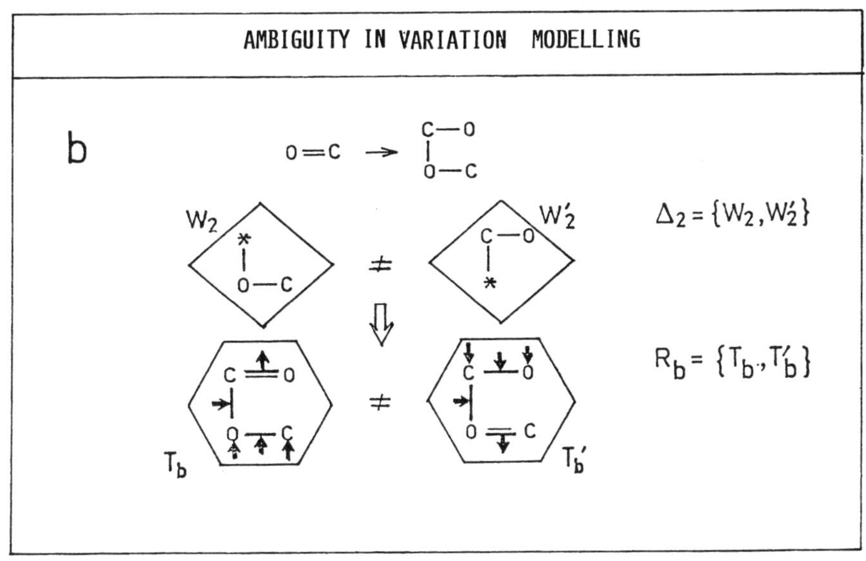

Figure 8.

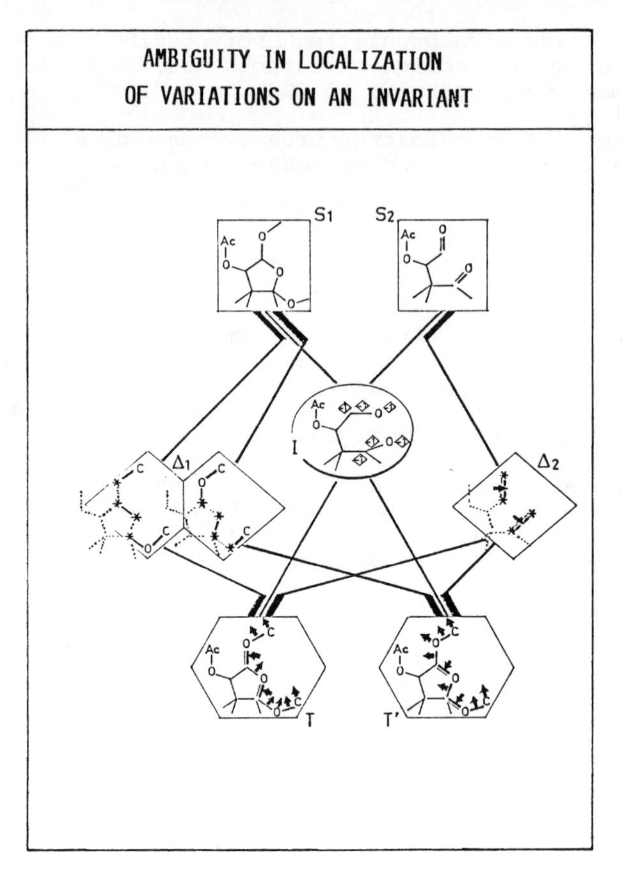

Figure 9.

2. TWO RESULTING SPECIES OF REACTION SUBSTRUCTURES

From the two structural models we have previously defined, we distinguish two species of reaction substructures. To define a substructure, one has to specify both syntax and semantics. Syntax deals with the structural type of the substructure, i.e., the proper mathematical formalism, while semantics gives the conditions under which a substructure of a given type 'is a substructure of' a chemical structure. Our aim in this paper is not to enumerate or to classify the different types of substructures conceivable inside each species. Thus we use substructure types formally equivalent to the structure type from which they are originated. Moreover we limit our scope to very simple substructural relationships related to partial subgraphs in ordinary graph theory. We then clearly differentiate the two species of reactional substructures by their consequences on induced populations of sub- and super-structures.

2.1 Pair of compound substructures

The substructure of a reaction modelled by a pair $P = (S_1, S_2)$ is a sub-pair constituted of a pair of usual compound substructures. It is defined from subgraphs of graphs representing each initial or final structure:

1) a chromatic graph g is a partial chromatic subgraph of G ($g < G$), if g derives from G by removing certain edges, or certain nodes and all edges adjacent to these nodes; the chromatic functions on g are simply the restrictions of corresponding functions on G;

2) a structure s is a sub-structure of S ($s < S$) if there exists $g \in s$ and $G \in S$ such that $g < G$;

3) a pair $p = (s_1, s_2)$ is a sub-pair of $P = (S_1, S_2)$ if $s_1 < S_1$ and $s_2 < S_2$: this is denoted $p < P$.

One can then explicitly or implicitly associate with each pair of structures the set of all its sub-structures, and the set of all its super-structures:

$\Gamma_<(P) = \{p \mid p < P\}$ is the population of sub-pairs of P,

whereas

$\Gamma_>(p) = \{P \mid P > p\}$ is the population of super-pairs of p.

2.2 Sub-transfostructures

The substructure of a reaction modelled by a transfostructure T is defined from a 'sub-transfograph' relation corresponding to the subgraph relation used for substructures:

1) a chromatic transfograph f is a partial chromatic sub-transfograph of F ($f < F$), if f is obtained from F by removing some edges, or by removing certain nodes and all edges adjacent to these nodes; here too the chromatic functions on f are simply the restrictions of corresponding functions on F;

2) a transfostructure t is a sub-transfostructure of T ($t < T$), if there exists $f \in t$ and $F \in T$ such that $f < F$.

Induced populations are then:

$\Gamma_<(T) = \{t \mid t < T\}$ the population of sub-transfostructures of T,

and

$\Gamma_>(t) = \{T \mid T > t\}$ the population of super-transfostructures of t.

2.3 Comparing the two species of substructures

Let g_1 and g_2 be the initial and final graphs derived from f, and G_1 and G_2 the initial and final graphs derived from F. If $f < F$, then $g_1 < G_1$ and $g_2 < G_2$. Thus if: $f \in t$ and $F \in T$; $(g_1, g_2) \in p$ and $(G_1, G_2) \in P$; $g_1 \in s_1$, $g_2 \in s_2$, $p = (s_1, s_2)$, $G_1 \in S_1$, $G_2 \in S_2$, $P = (S_1, S_2)$ and

$$t < T, \text{ then } p < P.$$

The sub-transfostructure relationship is thus more restrictive than the sub-pair relationship. As a consequence the population of pairs associated with transfostructures of $\Gamma_<(T)$ is included in $\Gamma_<(P)$, and the population of pairs associated with transfostructures of $\Gamma_>(t)$ is also included in $\Gamma_>(p)$.

2.4 Data base search with sub-reactions

According to the chosen model, two kinds of sub-reactions may be defined:

1) $r <_{C-O} R$, if the model of r is a sub-pair of the C-O model of R;

2) $r <_{R-O} R$, if the model of r is a sub-transfostructure of the R-O model of R.

As r is interpreted as a generic reaction, both of these sub-reaction relationships may be used to ask:

a) a generic query r in a specific Reaction Data Base $\{R_j\}$;

b) a specific query R in a generic Reaction Data Base $\{r_j\}$.

In both cases a) and b) the set of answers resulting from C-O query includes the set of answers resulting from R-O query. Whenever a difference occurs, it appears to be entirely composed of irrelevant answers (noise) according to current meaning of generic reactions. This is not contradicted with some misleading situations where queries using C-O model generate no noise.

For example, in the case of specific RDB search, the content of the Base (B) is confronted with $\Gamma_{>C-O}(r)$ and $\Gamma_{>R-O}(r)$. Two different situations can arise (Figure 11):

1) If the intersection of B and $\Gamma_{>C-O}(r)$ is entirely included in $\Gamma_{>R-O}(r)$ (or is empty), then retrieval is equivalent with C-O or with R-O (13); there is no noise with p-query.

2) Otherwise there is noise with p-query, and t-query should be used. This is generally the case as shown in our

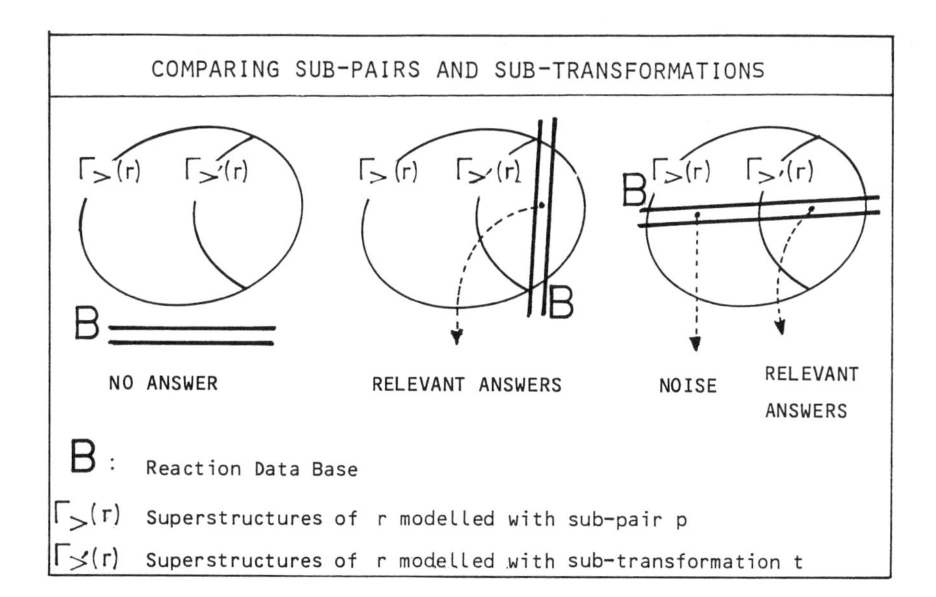

Figure 10.

examples of most frequently indexed methods in KETO-REACT (Figure 10).

We thus propose the $<_{R-O}$ as the sub-relationship best suited to confront generic reactions with specific ones. Intermediate solutions indicating broken and formed bonds (4), but without their relative localisation, are currently employed to reduce noise resulting from C-O query, but they do not reach the precision of R-O method.

Another kind of discrepancy in scope of substructural relationships is very interesting; it is linked to the possible parametrisation of sub-transfostructure relationships according to character (ℓ,i,e1 or e2). This property is illustrated on two important cases:

1) Generic interpretation may be limited to internal and ablation elements, leaving specific interpretation for loose and adjunction elements. This restriction describes generic retro-reactions as deterministic operators in backwards synthetic space generation (14).

2) In a similar way, allowing generic interpretation for internal elements only is frequent in structural change searches in specific RDB (Figure 11).

In conclusion, we think that a new generation of Reaction-Oriented Data Base Management Systems will arise from the graph theoretical progress in structural transformation modelling. The DARC Structural Management System has proved very efficient for compound handling; it takes into account the Reaction-Oriented (R-O) model for specific and generic reactions, and leads to sub-reaction relationships widely reducing noise in RDB retrieval. From the R-O model is derived a general language especially useful in Computer Aided Synthesis Design and Mechanism Elucidation Simulation (16).

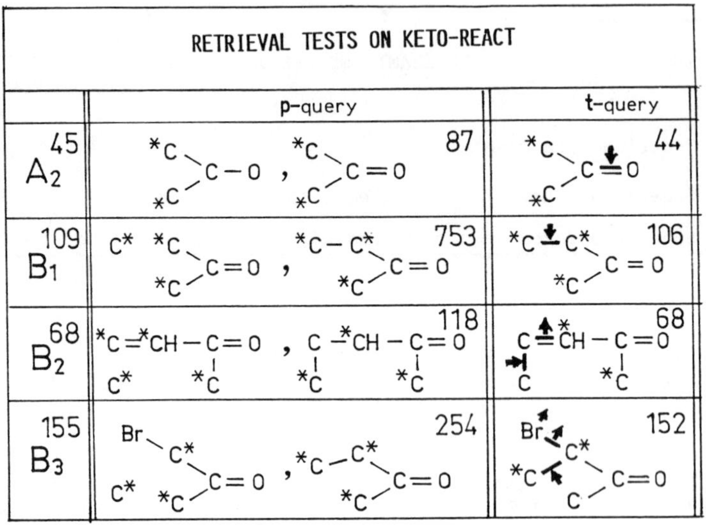

RETRIEVAL TESTS ON KETO-REACT

	p-query	t-query
A_2 45	(*C)(*C)C–O , (*C)(*C)C=O 87	(*C)(*C)C=O 44
B_1 109	C* (*C)(*C)C=O , (*C–C*)(*C)C=O 753	(*C–C*)(*C)C=O 106
B_2 68	*C=*CH–C=O (C*, *C) , C–*CH–C=O (*C, *C) 118	C=*CH–C=O (C, *C) 68
B_3 155	Br–C* C* (*C)C=O , (*C–C*)(*C)C=O 254	Br–C*, *C, (C)C=O 152

A_2 : Secondary alcohol oxydation

B_1 : Ketone alkylation

B_2 : 1-4 Addition on ethylenic ketone

B_3 : bromoketone alkylation

Figure 11.

REFERENCES

1. G. Gello. 'Question of Data Format in Organic Chemistry', J. Chem. Inf. Comp. Sci., 24, 249-254, 1984.

2. R. Picchiottino, G. Georgoulis, G. Sicouri, A. Panaye, J.-E. Dubois. 'DARC-SYNOPSYS: designing Specific Reaction Data Banks. Applications to KETO-REACT', J. Chem. Inf. Comp. Sci., 24, 241-249, 1984.

3. M. Bersohn, A. Esack. 'A Computer Representation of Synthetic Organic Reactions', Comp. Chem., 2, 103-107, 1976.

4. G. E. Vleduts. 'Concerning a System for Classifying and Encoding for Organic Reactions', Inf. Stor. Retr., 1, 117-146, 1963.

5. R. Attias. 'DARC Substructure Search System: a new Approach to Chemical Information', J. Chem. Inf. Comp. Sci., 23, 102-108, 1983.

6. J.-E. Dubois, A. Panaye, R. Picchiottino, G. Sicouri. 'DARC System: Structure of a Reaction Invariant', C. R. Acad. Sci., Paris, Series II, 295, 1081-1086, 1982.

7. J.-E. Dubois, G. Sicouri, Y. Sobel, R. Picchiottino. 'DARC System: Localised Operators and Co-Structures of a Reaction Invariant', C. R. Acad. Sci., Paris, Series II, 298, 525-530, 1984.

8. J.-E. Dubois. 'Ordered Chromatic Graph and Limited Environment Concept', in 'The Chemical Applications of Graph Theory', A. T. Balaban (ed.), Academic Press, London, pp.333-370, 1976.

9. G. Sicouri, Y. Sobel, R. Picchiottino, J.-E. Dubois. 'DARC System: Localising Variations onto the Reaction Invariant: the Transformation Structure Concept', C. R. Acad. Sci., Paris, Series II, 523-528, 1984.

10. G. Sicouri, Y. Sobel, R. Picchiottino, J.-E. Dubois. 'Representing, Coding and Handling Sets of Structured Entities' in 'The Role of Data in Scientific Progress', P. S. Glaeser (ed.), North-Holland, pp. 573-580, 1985.

11. J.-E. Dubois. 'Structural Organic Thinking and Computer Assistance in Synthesis and Correlation', Isr. J. Chem., 14, 17-23, 1975.

12. P. Chambert. 'Validation, Extraction and Completion of Structural Data from PRO-REACT', DEA, Paris 7 University, 1983.

13. Two subcases may be distinguished: either $\Gamma_{>R-O}(r)$ is strictly included in $\Gamma_{>C-O}(r)$, or $\Gamma_{>C-O}(r) = \Gamma_{>R-O}(r)$. This latter case is only encountered in more sophisticated formal types of substructures than those presented in this paper.

14. R. Picchiottino, G. Sicouri, J.-E. Dubois. 'DARC System: Transformational Relationships and Hyperstructures in Chemistry' in 'Data for Science and Technology', P. S. Glaeser (ed.), North Holland, pp. 229-334, 1983).

15. J.-E. Dubois, R. Picchiottino, G. Sicouri, Y. Sobel. 'DARC System: Block Relationships and Hyperstructures', C. R. Acad. Sci., Paris, Series I, 294, 251-256, 1982.

16. R. Picchiottino, G. Sicouri, J.-E. Dubois. 'DARC-SYNOPSYS Expert System: Production Rules in Organic Chemistry and Application to Synthesis Planning' in 'Computer Science and Data Banks', Z. Hippe (ed.), Polish Academy of Sciences, pp. 183-193, 1984.

18 Conference overview and closing remarks

W. A. Warr
Imperial Chemical Industries PLC Pharmaceutical Division

In his scene-setting paper, Dr Willett referred to the synthetic organic chemist practising his 'art in the midst of science'. As I looked around the audience at this meeting I was pleased to see a significant number of 'genuine chemists' amidst the familiar CSA/RSC faces.

Dr Pearson described for us the needs of the practising chemist. He described the academic's 'elegant' synthesis, the industrial chemist's need to make analogues efficiently and the process development chemist's desire to find a safe and relatively cheap reaction procedure. He also reminded us of the need to develop novel reactions, new reagents and new intermediates. He described some of the chemist's favourite tools for accessing reaction data and made the vital point that a chemist will not adapt to a computerised system unless that new tool is even more attractive than his existing favourites. After mentioning some of the special difficulties a computer system faces, he outlined what I will call the ten tips to have at your fingertips when evaluating any new system for reaction indexing:

1. There must be a sizable, good quality database.
2. It must be up-to-date.
3. Cosmetically, the system must be 'chemist-friendly'.
4. It must be easy to use.
5. Searching must be fast.
6. The system must handle stereochemistry, tautomers and, preferably, generic structures.
7. It should allow searching for selective reactions.
8. The signal to noise ratio should be high.

9. Data searching and display should be fast.

10. The system must always be available.

Perhaps we should have had these ten points on display during the sessions on computerised systems and my summary of them.

One of the longest established systems for reaction (and structure) documentation is the GREMAS system of IDC. A three letter fragment code is used for each carbon atom, and its environment, in the reaction centres of both starting material and product. An additional two characters encode carbon-carbon bond formation or cleavage, or ring formation or cleavage. The system can cope with several reaction sites per reaction and with multistage reactions. To reduce noise, a numbered carbon atom is traced through the various stages. The principal disadvantage of the system is the manual effort involved. Hoechst have recently been investigating full structure input, using the computer to generate the fragment codes. This has the additional advantage that the full structures of starting materials and products are then available for topological search.

The printed tools described by Dr Sinclair have no such advantages. He did not mention Current Chemical Reactions (CCR) from ISI or Methods in Organic Synthesis from the RSC, but it would be impossible to mention all the available printed tools in one lecture, without losing the attention of the audience. Dr Sinclair informs me that he does browse CCR to see if he has missed anything in browsing the primary literature.

Mr Bysouth concentrated most on two of the printed tools which have been computerised, that is Theilheimer as CRDS and CAS ONLINE. He said that it is not enough to know the structural nature of starting material and product - the chemist needs full preparative details. Does an online system do this better than the printed secondary publication? It does not matter how old a reference is, if it gives the chemist the recipe he needs. (I, myself, remember doing Fischer indole syntheses from last century's Annalen.) Older references are not covered by current online services. No doubt Mr Bysouth will find Beilstein online, back to 1830, useful when it appears. He pronounced himself in favour of a fully comprehensive reaction database. (In a later question time, Dr Oppenheim wondered whether chemists really need only one tried and tested recipe in most cases, rather than a comprehensive answer.) Mr Bysouth ended with a longer 'wish list' than Dr Pearson's - seventeen points this time, which increased to about twenty-three as question-time progressed. Maybe he is wise to ask for the sun, the moon and the stars in the hopes that the systems team will eventually deliver at least the moon.

After this we considered four commercially-available software packages for use in-house. Professor Wipke explored reactions in full technicolour with REACCS and M. Gay groped for them in the DARC. You can also browse for them in SYNLIB (many 'real' chemists certainly find this the most user-friendly method) or you can consult the Leeds ORACle. Interestingly, Dr Johnson had the presence of mind to take up the issue of Mr Bysouth's seventeen requirements and say that ORAC addressed most of them.

It was at one point suggested that my part in the conference should be to produce a 'Which?' guide to available software. This I refused to do (thus saddling myself with the more onerous task of the present paper) for the same reasons I refused to call my paper at the 1982 Conference in Exeter a 'Which?' guide. I quote (from myself):

> 'Different organisations have different user requirements and priorities and it is impossible to classify the software packages as 'good', 'bad' or 'best' except in terms of a combination of some subjective requirements and the importance placed upon each requirement.' (1,2)

I will restrict myself to making a comparison between Californian wine and French wine and let my listeners, and my readers, draw their own conclusions. The Californians market-research their product, design it with scientific care and dedicate half their personnel to the marketing of it. The result is an excellent and professional product at a very high price. The French set about things in a different way. The fine art which they practise produces a later product. We have to wait for it because it has to mature, but once it is ready it could well be better than the Californian one. Incidentally, British wine is beginning to have an impact in the market...

Having treated the vendors with a certain levity, I shall now discuss the academic contributions more seriously.

It was fitting that Dr Vladutz should start this session since he was probably the first to propound the reaction site hypothesis (3) in 1963. At this conference he reviewed the formalised definition of the reaction site concept and showed it can be used as a distance measure in a reaction network or map. Reaction distance networks can be used for 'intelligent' reaction retrieval - 'the system should read your mind' in Dr Vladutz's own words.

Dr Brandt has also studied the changed and unchanged parts of molecules and employed the mathematical model of constitutional chemistry first described by Dugundji and Ugi (4). His bond-electron matrices can be used for a hierarchical classification of reactions. A linear notation can also be derived from the canonicalisation rules. Having described his versatile reaction generator, Dr Brandt showed how the generated possibilities could be compared with actual reactions on a database.

Mr Cook followed with a paper on the automatic generation of keywords, to reduce the effort involved in inputting reactions to a system (and, more particularly, to ORAC).

Dr Bersohn believes that conventional reaction indexes do not convey the scope of reactions. He prefers to store a set of rules rather than a database of molecules. He regards the conversion of a ketone to a methylene group as one reaction which can be carried out under varying conditions, with related literature references. In Dr Bersohn's miniature system, the chemist inputs the structures of a product, and the reactant from which it is made. A retrosynthetic algorithm then works out if there is a reaction for making that product and under what conditions (e.g. interfering functional groups, steric hindrance) it is feasible. The output is a literature reference. One particularly interesting thing Dr Bersohn does, which few others bother with, is the checking of the input connection table for chemical invalidities such as cyclobutadiene or a double bond at the bridgehead in a six-membered ring.

I got the feeling that many of the audience did not appreciate the significance of Dr Bersohn's work. They should note that Professor Dubois cites it in the written version of his paper. The spoken version, given by M. Sicouri, was a model of mathematical ingenuity, above the heads, one suspects, of most of the audience. Two reaction management languages were compared, a reaction-oriented language and a compound-oriented language. In the former, the transformation is represented by a chromatic graph, whereas in the latter the reactant and product molecules are represented by chromatic graphs. Examples were taken from the KETO-REACT database.

The advantages of the researches of all these academic workers are that these respected teams are free to explore new possibilities, without the constraints imposed in industry, and cast new light on the complex problem of reaction indexing. There is seldom a single solution to a complex problem.

After lunch, in CONTRAST, we heard something completely different. CONTRAST is a 'bolt-on' to Pfizer's structure-handling system SOCRATES. I am told that SOCRATES is not an acronym. My encyclopaedia says the following of SOCRATES:

> 'He committed nothing to writing [Pfizer has not published] and our knowledge of him comes from the accounts of some of his pupils especially Plato [Bawden]. Socrates was concerned with probing the meaning and truth of ideas by a technique of question and answer [search algorithm] and by so doing, laid the foundations of systematic philosophical thought. At the same time, he called into question the institutions of state [the State of California?]. This made him enemies; he was brought to trial on a charge of corrupting the minds of the young and condemned to death ...'

Dr Bawden contrasted CONTRAST to SYNLIB (which has also been installed at Pfizer). The former is more of an 'information science system', whereas the latter is very chemist-oriented. SYNLIB seems to be renowned for its 'browsability' and Dr Bawden hopes to improve the browsability of CONTRAST, and build up its database, now that he has fully tested the initial software.

Next, Dr Finch had his chance to answer Mr Bysouth's criticism of CRDS. The system is not intended to be fully comprehensive and if evaluated in terms of the right criteria it is a very useful, working system. I remember Dr Finch raising the question 'Are graphics cost-effective?' at a meeting last year (5). He now seems to feel that graphics are inevitable, whatever his views on cost-effectiveness. Theilheimer will, of course, soon be available as a REACCS database from Molecular Design.

And finally, to the exponents of comprehensive databases, CAS and Beilstein.

The initial service from CAS will cover 300,000 reactions a year from about 100 core journals. In his paper in York, Dr Blower concentrated on indexing and input problems, on the program which will generate reaction sites and on quality control and final online edit.

When Beilstein online finally appears, it may be an even more comprehensive database than CAS (is this why CAS are concentrating on quality control?) since it will go back to 1830. This should help Mr Bysouth find those useful old references. Beilstein intend mounting an interesting combination of 'raw' and 'refined' data in order to be more up-to-date than they have been in the past yet to retain their famous image as producers of a high-quality, evaluated database. Eventually, they will be marketing software for use in-house. The second half of Dr Jochum's paper was the fascinating results of a huge market research survey on what people require from online database producers. Unfortunately, Dr Jochum will not be able to publish the results in detail.

So much for the conference overview, but what about the closing remarks?

Obviously, graphics are in, alas for Dr Finch.

Presumably, everyone will have to resort to CAS for his fully comprehensive searches, at least until Beilstein is mounted. The appearance of the high-quality Beilstein database online will be welcomed, but it will not offer reaction indexing as such.

How soon will major databases become available for mounting in-house, on laser disks or whatever? Certainly, selected parts of Beilstein will be available in the foreseeable future.

Meanwhile, what in-house solutions should we adopt for reaction searching? I have already said that I cannot advise you which software will suit you best. You may even find two packages are the answer. After all, Pfizer has both CONTRAST and SYNLIB. Would REACCS plus ORAC solve your problems (if you could afford both!)? Perhaps you may think DARC answers your problems because it gives you a CAS solution as well as an in-house one (although, presumably it will not offer CAS's newer databases). You, yourself, must decide whether to vote for compatibility, support, cheapness or whatever.

If you have not chosen already, you have the advantage that market competition is causing improvements in all the products. Attempts by any vendor, to monopolise a market by 'unfair' means could have no advantages for us, the victims. I have no objection, however, to someone seizing 80% of the reaction indexing market because his product is clearly the best in the eyes of a set of buyers given free choice amongst the competition.

Finally, we should bear in mind that an isolated reaction indexing package can never be as useful as one that is integrated with, say, LHASA, CAMEO, and in-house substructure-searching system (SOCRATES or whatever) and the Fine Chemical Directory (to give you access to available starting materials). This truly could be CAOS - computer-aided organic synthesis.

REFERENCES

1. Warr, W. A., 'Software for Storage and Retrieval of Chemical Structures', a paper given at the CSA Conference 'The Future of Chemical Documentation' at the University of Exeter, England, Sept. 6-10, 1982.

2. Kao, J., Day, V. and Watt, L., 'Experience in Developing an In-house Molecular Information and Modelling System', J. Chem. Inf. Comp. Sci., 25, 129-135, 1985 .

3. Vladutz, G., 'Concerning One System of Classification and Codification of Organic Reactions', Inf. Storage and Retrieval, 1, 117-146, 1963.

4. Dugundji, J., and Ugi, I., 'Algebraic Model of Constitutional Chemistry as a Basis for Chemical Computer Programs', Top. Curr. Chem., 39, 19-64, 1973.

5. Finch, A. F., 'Chemical Reactions Documentation Service - Ten Years On', paper given at the 188th National Meeting of the American Chemical Society, Philadelphia, Aug. 26-31, 1984.

Index

A

Automatic keyword generation 188-189, 200-201

B

Beilstein 32, 64, 165-169, 204, 261
Bond breakage and formation 14, 21, 57-58, 70, 124-125, 152-155, 162-164, 181, 189, 201, 202
Bond-electron matrices 223-225, 259
Browsing 85, 92, 104-114, 125, 261

C

Card indexes 20-21, 29
CAS ONLINE 32, 54-56, 64-67, 126, 146, 148, 258
Chemical Abstracts 20
Chemical Abstracts Chemical Substances Index 32, 33
Chemical Abstracts Formula Index 31
Chemical Abstracts Service Registry System 3, 32, 150
Chemical Abstracts Substances Index 34
Chemical graphics systems 26, 48-50, 61, 83-84, 87-88, 93-94, 120-121, 184-185, 194-197
Chemical nomenclature 3-4, 41, 170
Chemical Reactions Documentation Service 5, 11, 36-50, 56-59, 64-67, 258
Chemical structure and substructure searching 2, 13, 54, 57, 60, 82, 83, 99
Chromatic graphs 242-245
Compound-oriented models of reactions 241, 242-243, 249
Computer-aided synthesis design 28, 60, 78, 85-86, 119, 188, 223, 254, 262
Connection tables 8, 152, 171-173, 180, 182
CONTRAST 78-86, 260, 262
COUSIN 79, 118
Current Chemical Reactions 20, 258
Current Literature File 96, 114

D

DARC 87, 240, 253, 259
DataStar 64
Dictionary of Organic Compounds 30

E

Expert systems 118, 192

F

Failed reactions 150
Fine Chemicals Directory 97, 115, 262
Fragmentation systems 7, 8, 37, 68-74, 82-83, 258

I

Imperial Chemical Industries Pharmaceuticals Division 7
Index Chemicus 59
Internationale Dokumentationsgesellschaft für Chemie 7, 68-77, 258

J

Journal of Synthetic Methods 36-38, 46, 57, 96, 114

K

Keyword indexing 38, 40, 41, 55, 56-57, 186-187
Knowledge bases 119, 189

L

Literature searching 20, 28-35, 258
Lockheed DIALOG 64, 67

M

MACCS 93, 118
Maximal common subgraphs 12-13, 154-155, 162-164, 214-216
Mechanistic chemistry 7, 57
Molecular Design Limited 48, 59, 92, 96, 114, 261
Multi-step reactions 23, 70-75, 80, 96, 147, 258

N

Nearest neighbour searching 204-205, 218

O

Online searching 38-48, 51-67, 169, 258
ORAC 81, 184-201, 259
Organic Synthesis 4, 32, 92, 94, 98

P

Personal files 20
Pfizer Central Research (UK) 78, 85
Pharma Documentation Ring 7
Problems of data base input 149-152, 186

R

REACCS 48, 59, 92-117, 259
Reaction centres. See Reaction sites
Reaction conditions 3, 60, 88, 93, 123-124, 148
Reaction databases 1, 36-37, 54-61, 63, 79, 84, 87, 94-97, 114-115, 147, 158-159, 261
Reaction generators 226-227, 229-230, 236, 259
Reaction indexing 1-17, 76-77, 81-82, 85, 148-149, 185
Reaction Management System 87-91
Reaction matrices 208-209, 223-225
Reaction query types 2, 21-23, 52-54, 64-67, 75, 84, 88-90, 149
Reaction sites 5-14, 25, 70-71, 81-82, 88, 94, 104, 121, 152-155, 187, 206-208, 259
Reactiones Organicae 6
Reaction-oriented models of reactions 241, 243-246, 249, 253
Recreated reactions 151
Relationships between keywords 189-192, 198
Representation and searching of generic compounds and reactions 24, 54-55, 60, 61, 89, 121-122, 126
Requirements for reaction retrieval systems 26-27, 35, 61-62, 257-258
RINGDOC 60

S

Science Citation Index 32, 59, 67
Set-reduction 153
Similarity measures 204-205, 217-218
Smith, Kline and French 60, 119
SOCRATES 78-80, 83, 260, 263
Stereochemistry 22, 35, 48-49, 96, 99-103, 170-171, 180
STN International 146, 169
SYNLIB 60, 81, 85, 118-145, 259, 261, 262
Synthetic routes 18-19

T

Tautomerism 26, 88, 180
Theilheimer 5, 33, 36-38, 41, 43, 46, 57, 92, 95-96, 115, 261
Thesauri 185, 189
Transfostructures 243-247, 251

V

Validation and verification of data 147, 155-157, 178-180, 182, 189, 260

W

Wiswesser Line Notation 1, 9-12, 88